T0193398

ALSO BY RICHARD M. RESTAK

Brainscapes

The Modular Brain

Receptors

The Brain Has a Mind of Its Own

The Mind

The Infant Mind

The Brain

The Self Seekers

The Brain: The Last Frontier

Premeditated Man

OLDER

and

WISER

How to

Maintain Peak

Mental Ability for

As Long As You Live

RICHARD M. RESTAK, M.D.

Simon & Schuster

SIMON & SCHUSTER
Rockefeller Center
1230 Avenue of the Americas
New York, NY 10020

Copyright © 1997 by Richard M. Restak, M.D.
All rights reserved,
including the right of reproduction
in whole or in part in any form.

SIMON & SCHUSTER and colophon are registered trademarks
of Simon & Schuster Inc.

Designed by Liney Li

Illustrations by Jackie Aher

Manufactured in the United States of America

3   5   7   9   10   8   6   4   2

Library of Congress Cataloging-in-Publication Data
Restak, Richard M., date.
Older and wiser : how to maintain peak mental ability for as long
as you live / Richard M. Restak.
p.   cm.
Includes bibliographical references and index.
1. Cognition in old age.   2. Cognition—Age factors.   3. Human
information processing—Age factors.   4. Memory in old age.
5. Memory—Age factors.   I. Title.
BF724.85.C64R47   1997
155.67'13—dc21   97-14232   CIP
ISBN 978-1-4767-9288-0

At various places in this book, diet, levels of physical activity, medications and other chemicals are discussed in terms of improving brain function. Since the overall health effects of changing any of these factors may vary from person to person, the author and publisher suggest that a physician or other health professional be consulted prior to initiating any change in the reader's usual habits or lifestyle patterns.

ACKNOWLEDGMENTS

Thanks for interviews, discussions, suggestions, and helpful criticism are due to the following people, who were of invaluable help in the research-and-development phase of *Older and Wiser:* Carol Barnes, Art Buchwald, Neil Buckholtz, Robert Butler, Harriet Doerr, Charles Guggenheim, Olga Hirshhorn, Lois Mailou Jones, Robert Joynt, Robert Katzman, Claudia Kawas, Zaven Khachaturian, Nicholas Kittrie, Irving Kristol, Guy McKhann, Patricia Neal, Dionis Coffin Riggs, Albert "Allie" Ritzenberg, Chalmers M. Roberts, Robert Rubin, Daniel Schorr, Morris West, C. Vann Woodward.

To Carolyn,

with love and thanks

for your support

over the years of our marriage

CONTENTS

INTRODUCTION

Older and Wiser will give you the information you need in order to keep your brain functioning at its peak throughout your mature years. Everyone can benefit from reading this book, primarily individuals 40 and older. After age 40, the brain undergoes a series of changes that modern brain science, neuroscience, now understands sufficiently well that we can establish guidelines for preserving, even enhancing, our brain's performance during the later years of our lives.

In the past, the brain changes that take place over the life span were interpreted largely in negative terms: the brain was said to "deteriorate," or suffer a "decline," during the later years. "The brain loses fifty thousand or more cells every day of our lives" was one frequently voiced, overly pessimistic expression of this belief. But new discoveries about the adult brain are proving such pronouncements wrong.

Pessimism is yielding to well-founded optimism as brain scientists update their knowledge of how the brain ages. Since the brain is a dynamic, ever-changing organ, it should come as no surprise to learn that brain researchers are confirming what many experts have long suspected but couldn't prove: the brain of an older person is not inferior to that of a younger counterpart; instead, the brain of an 80-year-old is organized *differently* than that of a 35-year-old. In practical terms, this means the mature brain possesses strengths and assets that it lacked decades earlier.

Research is showing that our brain retains many of its capacities for normal, even superior, functioning well into our 80s and 90s; that aging does not involve the loss of large numbers of cells, particularly from the cerebral cortex, where our most elaborate thinking occurs; and that brain processing during our 60s and beyond doesn't follow a downward curve.

What's more, as we age, we retain considerable control over how our brain functions. And we can take practical steps to improve brain performance over our entire life span.

We aren't simply passive spectators who must adapt to inevitable cognitive decline. We can maintain healthy brain functioning by adopting certain lifestyle habits. Included among these are educational and learning activities, diet, exercise, general attitude, and attention to our physical and mental health. In *Older and Wiser* we will be discussing these factors as examples of how much we can do to positively influence our brain function during the later years of our lives.

As a start, we will briefly explore the nature of aging. We will attempt to answer the question: Why do we age? Everyday observation reveals that not all organisms age at the same rate. Why is it that our cocker spaniel is likely to live no more than about 15 years while our parrot can be expected to make it to 40 years or more? What mechanisms underlie aging? Indeed, why do we and all other living organisms age at all?

As we investigate aging we can examine a pivotal question: If we learn enough about aging, can we come up with some way of altering its course? Already there are indications that the aging process can be slowed up in animals by certain dietary regimens, especially severe caloric restriction. Do such approaches make sense for humans?

From there we will concentrate more specifically on our subject, the brain. What happens in our brain as we age? Normal brain aging will be contrasted with the aging that accompanies diseases such as Alzheimer's.

Normal brain aging is accompanied by the loss of brain cells in areas different from those found in Alzheimer's disease. Further, normal losses carry implications for such mental processes as concentration, quickness of response, and endurance. This is one of the reasons why an older person is at a slight disadvantage when competing with a younger person in tests of simple mental rapidity. But there are ways, which we will discuss, of compensating for this loss of speed. Besides, except under relatively rare and specific circumstances (such as piloting an airplane or responding to a surgical or medical emergency), speed is a relatively unimportant aspect of mental capacity.

The mature brain has its own ways of compensating for the inevitable slowing that accompanies aging. As a popular advertisement puts it, "There are times when deeper thinking is demanded." The mature brain specializes in "deeper thinking" via more sophisticated and comprehensive encompassing responses. In essence, this is the "wisdom" that people in former times attributed to their elders.

But what about those older individuals who aren't "wise" or even minimally mentally competent? One of the most feared consequences of aging is the possibility of coming down with Alzheimer's disease. The specter of this dread illness arises in almost everyone over 50 who momentarily forgets a name or cannot immediately retrieve some bit of memorized information.

Actually, Alzheimer's disease is only one of a host of illnesses that result in what neurologists refer to as *dementia,* a medical term for what is popularly referred to as senility. Dementing illnesses result from *diseases* affecting the brain. Many neuroscientists believe dementias are on the rise due to the effects of some as yet unidentified environmental factors. Others believe the incidence has remained the same and only seems higher because people are living longer. Years ago, when the life span was shorter, people died of other causes before they were old enough for the first signs of dementia to show up. Does this mean dementia may be an inevitable part of aging? If we live long enough, will all of us come down with Alzheimer's disease or some other form of dementia? We will explore two very different theories of why some people but not others are at risk for the illness.

In addition, we will discuss the steps we can take throughout our lifetime to lower the odds that we will come down later in life with Alzheimer's or other dementing illnesses. That phrase "lower the odds" is an apt one, incidentally. If you're old enough to be reading this book, you have learned by this time that there are no guarantees in life when it comes to health or longevity. The most that we can hope to do is shift the odds in our favor by making the right choices. *Older and Wiser* contains advice on how to do that.

Along with physical changes in its structure and composition as it ages, the mature brain undergoes transformations in certain mental functions as well. Psychologists refer to *cognition,* a word that serves as a synonym for various aspects of thinking. Cognition includes memory, intelligence, attention, language, problem solving, visual-spatial skills, and speed of response. Each of these mental faculties changes as the brain matures.

In some instances the process may involve a falloff from a prior standard of performance. Other cognitive processes change very little. A few actually improve. We will discuss how mental performance changes as we age, and suggest ways of maximizing our brain's performance.

Each of the mental (cognitive) processes will be taken up separately. Memory is the mental faculty most affected in aging, and it is also the faculty that arouses the most concern in all of us. Typically, many of us worry when we cannot immediately come up with a person's name. According to a recent survey of adults, almost 50 percent of respondents report worry about their memory. In the vast majority of instances their concern is unfounded. They are experiencing nothing more serious than age-associated memory impairment (AAMI), which isn't abnormal at all and is not a harbinger of Alzheimer's disease or any other dire brain impairment. We will discuss AAMI and how it can be distinguished from the memory loss accompanying Alzheimer's or other forms of dementia.

Not all brain disorders involve cognition. We are feeling and emotive as well as thinking creatures. As a result, we are prone to depression, particularly during late adulthood. Older people in the United States commit suicide at a higher rate than any other age group. Most kill themselves because they are depressed. While physical illnesses play a larger role in later-life depressions, not every person with a physical illness gets depressed or contemplates suicide. We will describe how depression can be recognized, treated, and in most instances cured or at least controlled.

Along with depression, the mature brain is also especially susceptible to anxiety, insomnia, and alcohol abuse. Each of these has specific effects in later life. Knowledge about how these disorders affect the mature brain can provide important information for wise, healthful management.

Problems in each of the areas mentioned so far (dementia, memory failure, depression, and so forth) characteristically evolve in a gradual manner. A person doesn't simply awaken one morning with a full-blown case of depression or dementia. But one illness—stroke—can occur in an instant. While very little could be done about strokes in the past, stroke is now treatable if diagnosed in time. Stroke almost always is preceded by early warning signs. If these signs are recognized and heeded, one can take steps to prevent the stroke. Our emphasis will be on prevention, early recognition, and treatment.

In contrast to the earlier ten books I have written on the human brain, *Older and Wiser* suggests practical applications for the new and exciting things neuroscientists are learning about this all-important organ as it ages. Throughout the book I will describe how we can influence for the better many of the age-associated brain changes.

Research is now providing proof that our longevity, along with our health during the second half of our lives, is directly related to how efficiently our brain is functioning. Indeed, if doctors are allowed only one piece of information about a physically healthy person of whatever age, and on that basis must estimate how long that person is likely to live, the doctor will ask about brain functioning.

Introduction

Successful and creative adaptation to advancing age is perhaps the most reliable measure of healthy brain functioning. My interest and curiosity about successful aging led me to interview many notable people on how they have continued to remain creative into their 70s and 80s. I present the advice of a dozen interviewees, advice we can apply to our own lives. On the basis of interviews with these "robust agers," as well as brain scientists specializing in the mature brain, I have compiled a list of thirty Pearls of wisdom, or suggestions about how to improve your brain's functioning and maintain that improvement.

ONE

1

AGING GENES AND

GENETIC CLOCKS

*B*rain aging is, of course, inseparable from general longevity. To state the obvious, the brain cannot live any longer than the organism as a whole. Before we talk about optimum brain performance, therefore, it's helpful to explore what scientists have learned about aging in general.

Just how long can we expect to live? That question is not as easily answered as it first appears. *The Guinness Book of World Records* credits Shigechiyo Izumi as the longest-lived person in modern times. But as with all such claims, legitimate doubt exists about Mr. Izumi's actual birthday. The oldest person of uncontested age is Mme. Jeanne Calment, of Arles, France. At the time of this writing she had celebrated her 121st birthday.

Judging from historical records, Mme. Calment's longevity

is about the limit. The maximum number of years that humans can live has not increased significantly over millions of years. But both the *average life span* (the average age actually achieved in the population) and the *life expectancy* (the number of years the average person can be expected to live) have shown impressive increases in this century, especially in developed countries.

If you were alive in 1900, you could be expected to live, on the average, until about 47 years of age. This was about the same length of life enjoyed by ancient Greek and Roman men born before 100 B.C. But longevity has increased more in the past fifty years than in the previous two thousand. Today the average life expectancy is about 71 years in men and 78 in women in the United States. What's more, life expectancy continues to increase. Between 1968 and 1979 it rose at a rate of one month per year for anyone over 50.

Longevity can also be considered in terms of population percentages. In 1900 only about 4 percent of the population exceeded 65 years of age. By 1980 the figure had risen to nearly 12 percent. It is predicted that by the year 2000 more than 15 percent of the population, about 35 million people, will be 65 years of age or older.

Until about 1970, this lengthening in average survival and the greater numbers of people living longer were the positive result of better prenatal care and the ready availability of antibiotics for the control of infectious diseases.

During the past two and a half decades, however, people are living longer because of attention to lifestyle changes. Fewer people are smoking. They are also paying more attention to their blood pressure and lowering their cholesterol through more healthy eating. Exercise, diet, weight control, and other health measures have helped lessen the incidence of heart attacks and stroke.

Can the average life span be further lengthened through applications of new technology? If we achieve an understanding of the scientific basis of aging, can we create a world filled with thousands if not millions of Mme. Calments? To answer that question we must first understand why and how we age.

• • •

Two major theories of aging compete for favor among scientists. The first, the *stochastic theory*, holds that aging is a random event. The *Oxford English Dictionary* defines "stochastic" as "randomly determined; that follows some random probability distribution or pattern, so that its behavior may be analyzed statistically but not predicted precisely."

Most health advice is based on stochastic principles. While people with normal blood pressure, moderate intake of alcohol, and total abstinence from tobacco tend to live longer than hypertensive, smoking alcohol abusers, we all know exceptions to the rule. While statisticians can predict the likelihood that one or another physical trait, habit, or behavior may lengthen or shorten longevity *on the average*, such reasoning does not make predictions possible in individual cases.

The second theory of aging, *programmed senescence*, is based on the belief that aging results from a genetically programmed set of events. According to this theory, a creature's life span depends on a series of processes beginning at conception and extending over a given number of years. Implicit in this theory is the idea that longevity can be prolonged if we discover the nature and functioning of the "biological clock" that determines how long we will live.

These two theories are not mutually exclusive; each is at least partially true. We know programmed senescence exists because each species has a distinct life expectancy, which can range from days to centuries. Just knowing the species to which an organism belongs enables us to make a fairly accurate guess as to how long a member of that species is likely to live. But at the individual level—estimating the likely longevity of a particular member of a species—the stochastic principle applies. Some people, due to accidents, illnesses, or any of the other unpredictable and uncontrollable aspects of life, will live less than the average, no matter how advanced the average life expectancy may become. On the other hand, increasing numbers of Mme. Calments are predicted in the twenty-first century.

Scientific proof for programmed senescence comes from the study of longevity in individual body cells rather than just the organism as a whole. In theory, longevity is determined in each species principally by the life span of the particular organism's constituent cells. The most thorough student of cell longevity in the laboratory was Leonard Hayflick, who in the 1970s spread a layer of fibroblasts (skin cells) in a culture dish under sterile conditions. When the cells multiplied sufficiently to form a single layer of cells covering the dish, Hayflick removed a few of the cells and transferred them to a second dish. The cells once again grew until this dish was filled as well. At about the thirty-fifth passage of the cell lines, the rate of division began to slow, and by the fiftieth division, the cells stopped dividing altogether.

Hayflick also found that cells taken from older people doubled fewer times than cells from embryos and younger people. In fact, the age of the donor could be correctly inferred from observing the number of cell doublings. This phenomenon, dubbed the *Hayflick limit*, also varies according to the life span of the species selected. Fetal cells from the mouse (life span 3 years) divide about fifteen times before reaching the limit, while cells from the Galapagos turtle (life span approaching 200 years) divide about ninety times. Hayflick's research provides evidence for the existence of a "genetic clock": an outer limit beyond which the organism cannot survive.

Along with many other scientists, Hayflick explains the differences in life span between cells as resulting from genetic *programs* within the cells. In the fetal stage of development, all of the genes are operating alike in each body cell. With further development, different genes turn on in different cells, and as a result, cells begin to evolve toward distinctive structures and functions, such as brain, stomach, liver, and lung.

Selective genetic expression explains also why cells throughout the body have widely varying life spans. Cells in the skin, bone marrow, and intestine, for instance, are continuously replaced over our lifetime. Brain cells, in contrast to all other cells in the body, do not generally divide once formed. Thus the brain is a very different organ from all of the others. Moreover, aging of the brain, both as a whole and in terms of its constituent cells (the neurons), carries with it specific consequences not only for our longevity but also for the quality of our lives. This preeminent importance of the brain in comparison to other organs and body cells provides one reason for caution in interpreting Hayflick's results.

The outer limits of cell multiplication discovered by Hayflick pertain to only one type of cell and may bear little relevance to the life span of the total organism. Nor does the research bear any relevance at all to brain cells, which, since they do not divide, do not suffer from a Hayflick limit. It's an interesting thought that, after an initial period of winnowing of our brain cells during infancy, we are in possession, by early childhood, of as many brain cells as we are ever going to have.

• • •

I mentioned the action of genes. Genetics plays a major role in determining how long the brain can be expected to function at peak performance. Consequently, we should understand a few important things about genes and their relationship to longevity.

A belief in genes is about as close as some of us come to believing in the invisible. We don't see genes but must infer their presence and the laws governing them from the effects they produce: flower color in plants or eye color in ourselves. Genes and the life span are intimately linked. Among our mammalian relatives only whales, orangutans, and elephants live, like us, into their 70s. Dogs rarely live much beyond 15 years. But among birds, pet parrots have served as heirlooms handed down over several generations. While good care and favorable circumstances can extend these figures slightly, genes set the outer limits: the world is still waiting to encounter the first 90-year-old-dog.

These differences in longevity between species are based on the action of genes: coded segments of DNA (deoxyribonucleic acid). Genes are DNA molecules that measure only a few hundred atoms, so small they can be seen only with the aid of powerful electron microscopes. They are strung like delicate pearls on a string extending along the twenty-three pairs of chromosomes within the nucleus of almost every human cell. (The exception are the red blood cells, which lose their nuclei to gain additional room for their cargo of oxygen.)

Despite their diminutive size, genes are the basis of life, a neat proof that small really can be better. They contain the master plan directing development and longevity of an entire plant or animal from a single fertilized egg. Genes form the basis for our every heartbeat, breath, movement, intestinal contraction, and thought. They also serve as a bridge to succeeding generations by directing the formation of sperm and egg and then hitching a ride for a unique form of time travel: transferring the blueprints of life into the next generation and the next, in the process we refer to as heredity.

Despite vast, easily observable physical and behavioral differences between species, their DNA varies only slightly. You cannot chemically distinguish between mouse DNA, horse DNA, and human DNA. The difference between a senator and his pet bearded collie lies in the way four chemicals are lined in a row. This quartet of chemicals, or nucleotide bases, as they are referred to, is usually named by the first letters of the chemicals' names: G,C,T,A, or guanine, cytosine, thymine, and adenine. All of life can be expressed in some variation of these four chemicals, their sequence and number varying from gene to gene.

Genes are often referred to as the blueprints of life. A better analogy would be a recipe with the ingredients listed in the order they must be mixed in order to turn on the chemical machinery necessary to produce a particular cell. From just these four chemicals, billions of combinations are possible, more than enough to allow for the variety of living organisms.

Indeed the entire human race throughout all of history represents fewer than 1 percent of the possible human nucleotide arrangements. One percent is also the difference between our own genetic profile and that of chimpanzees, our closest relatives. While this may seem like a small amount, 1 percent of the three billion or so DNA bases in our genes comes to 30 million differences between ourselves and the chimps. To make these differences even more intriguing, only about 1 percent are of any consequence in explaining why we and not chimps have gone on to develop language and create civilizations. So far scientists haven't a clue about which genes comprise that critical 1 percent.

• • •

With the discovery of genes and the different gene combinations characteristic of different species, scientists speculated about the existence of an "aging gene": a segment of DNA that accounted for the longevity differences between species. The search for an aging gene in plants and animals has led to the discovery of longevity-related genes in fruit flies and tiny worms (nematodes). And yeast cells possess fourteen genes that appear to be related to aging. If any one of these fourteen genes is activated—overexpressed or underexpressed—the life span of the yeast is increased or decreased by 30 percent. This LAG-1 (longevity assurance gene) works by influencing, through an unknown process, the number of cell divisions in the yeast. Is there a LAG-1 gene in humans, and if turned on, will it promote human longevity? The closest thing to an aging gene in humans occurs in the rare illness *progeria*.

Imagine yourself stepping into a fairy tale and encountering a 10-year-old child transformed by a wicked fairy godmother into a wizened old man. Despite having the height and weight expected in a 10-year-old, this magically altered child has lost most of his hair, suffers from high blood pressure, atherosclerosis, and other age-associated symptoms. And despite the appearance of old age, this person talks and acts like a typical early adolescent of normal intelligence.

Unfortunately, this scenario is not imaginary. Progeria is the name for premature and telescoped aging, and it can occur in two forms. The symptoms of the first, so far limited to less than thirty patients worldwide, consist of severely retarded growth, abnormal skeletal development, absence of sexual development, atherosclerosis, and usually death by 12 years of age. The second form of the illness is only slightly less unfortunate. Those afflicted with this form of the illness don't start to visibly age until adolescence. Then they stop growing and develop atherosclerosis and the skin changes associated with aging. In most instances, progeroids retain perfectly normal brain function throughout their short lives. Most are dead in their 30s.

A few years ago scientists pointed to this remarkable and fortunately rare genetic disease of progeria as proof for an aging gene. Nobody believes that now. For one thing, progeroids do not exhibit all of the traits seen in the elderly. Rather, progeria is a kind of pastiche of old age—some melodies are present, many others conspicuously absent. In addition, no single gene or series of genes has been found responsible for the illness. Normal aging, too, is probably influenced by a host of genetic factors.

With the exception of progeria, no evidence exists in our own species for alteration of the onset time for the genes responsible for aging. Normally, aging occurs as the result of the influence of a large number of genes combining together.

Aging is thus more like height, weight, and intelligence rather than hair color. Nonetheless, the possibility remains of influencing one or more of the genes responsible for setting off the cascade of chemical events that we call aging.

A moment's thought about the matter, however, casts serious doubt on the proposition that longevity can be increased simply by turning on genes and reinitiating cell multiplication. If immortality is linked with the capacity to multiply endlessly, one need look no further than to a cancer cell for proof of our body cells' potential for immortality. When cancer cells are cultured and transferred from one laboratory dish to another, they will continue to divide for as long a period as the experimenter is willing to harvest them. This suggests that discovering the factors underlying cancer-cell multiplication may help in the design of research aimed at increasing our understanding of aging. It also suggests that at this stage in our knowledge, the fate of individual body cells is of only limited help in our understanding of the overall aging process.

• • •

In the 1930s Clive McCay, a Cornell University scientist intent on discovering the secrets of longevity, observed that if he fed rats a diet about 40 percent of their normal intake, they lived a year longer, a life span of 4 years instead of the usual 3. The rats were supplied with all of the required amino acids, vitamins, and trace elements. They were deficient only in calories.

Since McCay's original observations, caloric restriction (a balanced reduction of the protein, carbohydrate, and fat content of food without reduction of the nutrient content) has been shown to slow aging in creatures as diverse as single-celled protozoans, roundworms, fruit flies, fish, and rodents. But this requires significant caloric reduction. Restriction of fats, carbohydrates, or proteins does not increase the life span unless calories are reduced from 30 to 50 percent. An animal placed on a diet containing 35 percent fewer calories can be expected to live 35 percent longer.

Animals raised on caloric restriction not only live longer but also live healthier. Various signs of aging are halted or delayed with caloric restriction. In addition, animals on caloric restriction develop fewer tumors.

These multiple benefits of caloric restriction led neuroscientists to believe only a few years ago that the primary aging process itself might be positively affected by reducing the number of calories ingested. But as a practical point, attempts to reproduce caloric restriction in humans are complicated by the length of the human life span. Because we live so long in comparison with other creatures, confirmatory evidence of the benefits of caloric restriction would require observations carried out over a hundred years. And there is another practical difficulty: How do you reliably distinguish between a *causal* and a *correlative* association?

For instance, graying of the hair is a reliable correlate of aging. But few people would assent to the proposition that graying of the hair is the cause of aging. Rather, gray hair correlates with aging. While this proposition seems a matter of common sense, other causal-versus-correlative decisions are far less simply resolved. For example, normal aging is accompanied by a general increase in the "stickiness" of the blood platelets along with an increase in diseases of the arteries. Do these physiological processes correspond to the graying of our hair—correlative accompaniments to aging—or are they in fact a *reason* why we age? As a possible causative mechanism, both platelet stickiness and narrowing of our arteries could serve to decrease the blood supply to an important area of the brain and thereby trigger a cascade of chemical events that result in aging.

One way of getting around these difficulties in discerning cause from correlation would be to identify reliable biomarkers, physical changes that serve as indicators of how healthily the aging process is progressing. But at the moment, no reliable biomarker exists. Everyday observation confirms this: from appearance alone it's obvious that not all people age at the same rate or look the same at a given age.

Until scientists can develop biological standards, or biomarkers, no definitive statements can be made about alleged positive effects on longevity brought about by caloric restriction, drugs, vitamins, or any other interventions. As things now stand in longevity research *in humans,* there is no alternative to keeping careful records and letting scientists sort out the results a hundred years from now.

According to Richard Weindruch, director of the Institute on Aging of the University of Wisconsin, Madison, it may take a decade or more before scientists will know whether caloric restriction might reap the same health benefits for people as it does for a variety of other creatures. In the meantime he suggests that a reasonable caloric-restriction regimen would consist of a daily consumption of roughly one gram of protein and no more than about half a gram of fat for each kilogram (2.2 pounds) of current body weight. For a 160-pound person, that would break down to a daily intake of about eighty grams of protein and forty grams of fat. Such a diet should be professionally monitored to guarantee that it contains complex carbohydrates (long-chain sugars such as are found in fruits and vegetables) along with all essential vitamins and nutrients.

Why and how does caloric restriction exert such healthy effects on aging? Only one explanation has so far stood the test of time.

In the 1950s Denham Harman of the University of Nebraska Medical School became curious about the high chemical reactivity of a class of molecules called *free radicals*. A free radical has nothing to do with the counterculture or a political affiliation but is a fragment of a molecule or atom containing at least one unpaired, or odd, electron. Normally, atoms consist only of paired electrons. Unpaired electrons are unstable. An atom with an unpaired electron does everything possible to attract other electrons in order to return to the paired state. In a power grab at the atomic level, free radicals steal electrons from other atoms or molecules. This starts a chain reaction in which atoms and molecules steal electrons from one another. The whole process is like the game of musical chairs, only in this case it's electrons and not chairs that are in short supply and are coveted by the molecules that want to stay in the game of life.

In order to achieve wholeness, free radicals aren't particular about where they obtain their electrons. A favorite ploy is to steal them from vital molecules that form the outer membrane of the neuron and of other body cells. They may also attack the proteins that compose the neuron's internal structural components. Eventually the power grab gets really serious and the DNA molecules in the nucleus at the center of the neuron are attacked. Neuroscientists believe that the ensuing damage to DNA is at least partly responsible for aging. Harman was among the first to propose that aging was caused by free-radical activity damaging and modifying DNA.

Thanks to increasing knowledge of the mechanisms of cell growth and development, scientists now believe that free radicals exert their maximum damage on *mitochondria,* tiny intracellular structures located within the cytoplasm surrounding the cell nucleus. Mitochondria serve as power generators for the cell's activities.

The effect of free radicals is similar to what happens with the explosion of a bomb or land mine: damage is maximal in the immediate vicinity of its detonation, in this instance the inner mitochondrial membrane.

Eventually a self-perpetuating cycle of destruction arises. As a result of mitochondrial damage, fewer energy-producing substances are produced, and this increases the production of free radicals. Additional free radicals wreak additional damage on the mitochondrial components. Eventually the body as a whole becomes less able to sustain or renew itself against the stresses of the environment—indeed, this is a functional definition of the aging process.

"It is likely that the maximum life span of a species is largely determined by the rate of mitochondrial damage inflicted by free radicals arising in the mitochondria in the course of normal metabolism," according to Denham Harman. "Measures that slow initiation of damaging free-radical reactions by mitochondria may increase both the maximum and functional life spans. Prospects are good that such measures will be found."

The proposal that aging results from free-radical damage to the mitochondria carries with it certain implications regarding diet. The body can employ antioxidants such as vitamins E and C as well as enzymes such as superoxide dismutase (SOD) to soak up free radicals and thereby slow the chain reaction. SOD converts oxygen radicals into hydrogen peroxide, which is then converted by a second enzyme into the normal constituents of the body: oxygen and water. SOD levels are highest among longer-lived animals. In fact, according to National Institute on Aging research chemist Richard Cutler, the varying life spans of different animals depend upon their ability to repair damaged DNA via the action of SOD.

According to Cutler, the amount of SOD in a particular animal is related to its metabolic rate, that is, how fast it burns up oxygen. Mammals with high metabolic rates, like mice, live a maximum of 4 years, while elephants, which consume far less oxygen proportionately, can live into their 70s.

Free radicals can also be neutralized by certain vitamins, which act as antioxidants. The older you are, the lower your plasma levels of vitamins C and beta-carotene (a precursor form of vitamin A that is favored since vitamin A can be toxic in high quantities). The mature brain requires more than the minimum daily requirements (MDRs) for vitamins C and beta-carotene. But how much more? At this point no one is certain, with most experts now favoring an increased intake of fruits and vegetables rather than of vitamin supplements.

Studies done at Oxford University and elsewhere indicate that vitamin supplements cannot substitute for a diet rich in fruits and vegetables. This should pose little problem for those of us who are following the latest USDA advice to eat five or more servings a day of fruits and vegetables. In order to meet the two-hundred-milligram level of vitamin C, a person needs to consume only half a cup of chopped, cooked broccoli, six ounces of cranberry juice, half a cup of orange juice, and one kiwi fruit. The list that follows itemizes some commonly available vegetables and fruits along with their vitamin C content.

VEGETABLES

Vitamin C	Food	Serving
49.0 mg	*broccoli*	*½ cup, chopped, cooked*
48.4 mg	*brussels sprouts*	*½ cup, cooked*
34.4 mg	*cauliflower*	*½ cup, cooked*
26.7 mg	*kale*	*½ cup, chopped, cooked*
38.8 mg	*peas with edible pods*	*½ cup, cooked*
94.7 mg	*sweet pepper*	*1 raw pepper*
26.1 mg	*potato, flesh and skin*	*7 ounces, baked*
26.3 mg	*rutabaga*	*½ cup, cooked, mashed*
24.6 mg	*sweet potato*	*½ cup, baked, mashed*
21.6 mg	*tomato*	*4½ ounces, raw*
25.1 mg	*tomato*	*½ cup, cooked*
22.3 mg	*tomato juice*	*½ cup*

FRUITS

Vitamin C	Food	Serving
33.7 mg	*cantaloupe*	*½ cup, cubes*
67.3 mg	*cranberry juice*	*6 ounces*
41.3 mg	*grapefruit*	*half a fruit*
79.1 mg	*grapefruit sections*	*1 cup with juice*
36.0 mg	*grapefruit juice*	*½ cup, canned*
29.8 mg	*grape juice*	*½ cup, frozen concentrate*
165.0 mg	*guava*	*3 ounces*
21.0 mg	*honeydew*	*½ cup, cubes*
74.5 mg	*kiwi*	*1 medium fruit*
22.8 mg	*mango*	*½ cup, sliced*
80.3 mg	*navel orange*	*5 ounces*
62.0 mg	*orange juice*	*½ cup, fresh*
43.0 mg	*orange juice*	*½ cup, canned*
48.4 mg	*orange juice*	*½ cup, concentrate*
43.2 mg	*papaya*	*½ cup, cubes*
30.8 mg	*raspberries*	*½ cup, fresh*
42.2 mg	*strawberries*	*½ cup, fresh*
25.9 mg	*tangerine*	*2½ ounces*

Although no one is certain why only natural foods and not supplements serve as longevity-promoting factors, one possibility is suggested by Dr. Ronald Krause, director of the Division of Molecular Medicine at Berkeley National Laboratory at the University of California, Berkeley:

One speculation is that perhaps vitamins are only effective when combined with one or more additional plant chemicals; if the vitamins are taken in the absence of these presently unknown substances the vitamins prove ineffective. There are thousands of plant chemicals that may contribute to disease prevention. We can justify quite strongly the importance of trying to get these substances from real food.

But whatever the reason, it seems fair to conclude that you can't buy good health and longevity in a health food store, but you might be able to do so if you make the right choices in your supermarket. You might prefer to hedge your bets by combining a vitamin-rich diet with some supplements. A reasonable regimen is given in the "Pearls" section at the end of the book.

• • •

No discussion of aging would be complete without reference to the amazing brain research on *growth hormone* carried out by the late Daniel Rudman, endocrinologist and professor of medicine at the Medical College of Wisconsin in Milwaukee.

Under normal conditions, growth hormone is released from the pituitary, a pea-sized gland deep within the brain that secretes hormones regulating the growth and activity of several other hormone-secreting glands, such as the thyroid.

Growth hormone is released in a pulsatile fashion. The release can be increased by vigorous exercise (walking and even jogging aren't sufficient) or by low blood-sugar levels due to any cause, including fasting, sleep, trauma, and the actions of certain drugs.

Growth hormone works by stimulating body repair of protein synthesis and cell division. Under its influence, bodily organs increase in size. In fact, if growth hormone reaches excessive levels due to a tumor in the pituitary, the elevated growth hormone causes acromegaly, a condition marked by enlargement and thickening in the skin, soft tissue, and bones, especially those of the face, hands, and feet. Too little growth hormone also leads to serious consequences. A deficiency in the childhood growing years leads to dwarfism, severely stunted growth. Administration of growth hormone to children suffering from dwarfism can reverse this process.

Rudman's interest in growth hormone stemmed from the fact that, as we age, our pituitary drastically reduces the circulating levels of growth hormone. The level for a 60-year-old is about 10 percent of the average level for 40-year-olds. This reduction is thought to correspond to the shrinkage in size that occurs with age in most body organs. Liver, lungs, kidneys, spleen, and muscle—organs that make up the lion's share of what is referred to as lean body mass—all get smaller. With these changes in mind, Rudman asked what might happen if an older person were given regular injections of growth hormone, injections below the level leading to acromegaly but just enough to reverse some of these age-associated shrinkages. Linked with that question was another related one: As the growth hormone injected into older volunteers increased the size of their body organs—restoring them to the size of decades earlier—would the person experience any reversing of the aging process?

In the mid-1980s Rudman administered regular injections of synthesized human growth hormone (HGH) into elderly volunteers. Within three months the volunteers noted increases in muscle mass and started to report that their clothes and shoes fit too tightly. This was the result of an increase in muscle mass rather than a gain in weight. The volunteers showed a 15 to 20 percent increase in the girth of their arms and legs, a 10 percent increase in liver and skin thickness, and a 22 percent increase in the spleen—the body's primary mediator of the immune response.

After only six months of treatment, Rudman's patients had regained somewhere between fifteen and twenty years of their youth in terms of appearance. The volunteers also reported feeling younger than they had in years. Nor was the perception merely psychological. One man married to a woman fifteen years his junior (she was 50 years old) reported that for the first time in years he experienced no difficulties keeping pace with his wife's demanding physical routine (she routinely arose at 5 A.M. and continued a whirlwind pace for the rest of the day).

In 1991 Rudman began weaning his subjects off the growth hormone. Within weeks, his subjects' youthful appearance and endurance faded. Everyone looked and acted once again his rightful age. With the decrease in energy, the subjects reverted to their old habits.

Although scientists had no explanation why growth hormone had exerted such an age-ameliorating effect, Rudman speculated that the hormone had reactivated dormant genes that years earlier had served to keep tissues and organs youthful.

Rudman's research continues to fascinate researchers. Growth hormone has come the closest to achieving what was promised by the Spanish explorer Ponce de León's mythic Fountain of Youth. Rudman's patients had bartered six months of old age in return for twenty years of regained youth .

But in 1996 scientists took another look at growth hormone as an aging retardant, and this time the results were far less impressive. Physicians at the University of California, San Francisco, conducted a growth-hormone study involving fifty-two men aged 70 or older. Half the subjects took growth hormone while the other half took a placebo. Neither the subjects nor the doctors knew who was receiving growth hormone. After six months the subjects on growth hormone showed changes in body composition predicted by the Rudman study six years earlier: increases in lean body mass (made up of muscles, internal organs, and body water) along with a decrease in body fat.

Unfortunately, these anatomical changes were not accompanied by functional changes such as increases in muscle strength, endurance, or mental acuity or positive mood changes. In addition, many complained of troublesome side effects such as leg and ankle swelling, aching joints, and stiffness in the hands. These side effects were sufficiently bothersome that over a quarter of the subjects had to reduce their dosage of the hormone. None of the subjects reported the Fountain of Youth experiences reported by Rudman.

Although no one knows for certain why the San Francisco

study turned out so differently—and Rudman is no longer alive (he died in 1994 at age 67)—the most likely explanation involves the differences in how the two studies were carried out. In Rudman's study, the subjects knew what they were getting, and their improvement may have been due to a placebo effect. Rudman's widow, also a researcher and a co-author of the 1990 study, agrees with this interpretation: "I think those men were expecting to feel better and they did. They became euphoric. Sometimes it's the psychological effect of expecting to feel better."

The growth-hormone studies illustrate the dangers of relying on subjective, testimonial evidence. A person can feel better and younger for at least brief periods of time for a multitude of reasons. And positive physical changes do not necessarily equate with enhanced functioning or greater longevity. Finally, the desire to halt the aging process can take on the quality of an obsession, a character trait I have observed not only in experimental subjects but in many of the researchers as well. For these reasons I encourage you to cultivate a healthy skepticism and remember this growth-hormone study later in the book when we take up other drugs and chemicals that are reputed to extend the life span.

• • •

In summary, two theories exist about aging. According to the first, aging is a programmed process brought on by one or more aging genes and controlled by an intrinsic clock. These aging genes restrict life expectancy by limiting the capacity of body cells to reproduce themselves. According to the second theory, aging is a consequence of the accumulation of errors brought about by free radicals and other oxidants. This the-

ory holds the hope for treatments aimed at reversing or preventing free-radical damage. Most likely an all-encompassing
theory of aging will incorporate elements of both theories.

This much we are certain of: over our lifetime all the cells
in our body are exposed to many DNA-damaging agents, including forms of intense radiation and highly reactive and
toxic products resulting from cellular metabolism. (These alterations are usually not inherited, because the resulting mutations in the reproductive, sperm and egg, cells effectively
eliminate these damaged cells as candidates for successful
fertilization.)

If unchecked, these various harmful influences may damage our DNA in cells throughout the body. This leads to the
introduction of "errors"—mutations and other damage to the
genetic programs encoded in our DNA. To counter this constant attack on our genetic material, our body cells have developed various protective mechanisms to repair the DNA
damage. These inherent protective measures against free-
radical and other damage can be aided by proper diet and, as
we shall see later, perhaps certain free-radical-neutralizing
drugs. It is especially important to prevent the accumulation
of genetic damage in brain cells, since they do not divide and
therefore must remain functional throughout our life span.
While the body can replace liver cells or cells lining the stomach or any other organ, neurons (brain cells) cannot be replaced. This leads to an operating principle: our best chance
of remaining mentally and emotionally intact throughout our
lives is to do everything we can to ensure normal brain functioning.

Our brain's level of functioning becomes particularly important later in life, when it, along with all of the other organs

in the body, must contend with the changes brought about by normal aging. But despite the brain's importance in aging, few of us know much about how the brain changes over the years from young adulthood into middle age and beyond. What knowledge we do possess is often mixed with misinformation. How much do you know about the way your brain ages?

1. Every day we lose thousands of brain cells as an inevitable consequence of aging.

True   False

2. What is the leading cause of disability in the United States?

   a. heart disease

   b. cancer

   c. brain and nervous system diseases

3. If you live long enough, the odds are you will develop dementia (formerly referred to as senility).

True   False

4. Memory failures are part of the aging process, and not much can be done about them.

True   False

5. If one of your parents suffered from Alzheimer's disease, the odds are you will come down with the illness too.

True   False

6. Creativity rarely occurs late in life, because the brain at that point is fixed in its patterns and responses.

True   False

7. Brain-associated disorders/diseases associated with aging include

    a. loneliness

    b. depression

    c. insomnia

    d. anxiety

    e. all of the above

    f. all of the above except *a*

8. The older you are, the less likely it is that you can learn new information.

    True   False

9. Alcohol is harmful to the brain and should either not be consumed at all or severely limited.

    True   False

10. The older you get, the less time you sleep.

    True   False

ANSWERS

1. *False.* The degree of nerve-cell loss differs tremendously from one person to another. There is no truth to such statements as "Every day a person loses fifty thousand or more brain cells."

2. *Brain and nervous system disorders* are the leading cause of disability and account for more hospitalizations and chronic care than almost all other diseases combined. Cancer and heart disease are the leading cause of death.

3. *False.* Dementia is a disease and a sign that the brain is not normal. A healthy brain shows none of the signs of dementia and will continue to function normally as long as we live.

4. *False.* It's true most older people are not as quick at retrieving information as they once were, but otherwise their memory is no different from that of a younger person.

5. *False.* Genetic tests are available that provide a rough estimation of the odds that a person may be at risk for Alzheimer's disease, but accurate predictions are not possible in individual instances.

6. *False.* The brain is constantly reorganizing itself on the basis of a person's life experiences. With creative people this reorganization continues across the entire life span. In addition, there are numerous instances of creativity beginning in the sixth decade or later.

7. *All of the above except* a. Loneliness is not a disease but rather the foremost challenge individuals face in their later years. How successfully an older person manages this challenge will determine how well his or her brain will continue to function.

8. *False.* While it's true that learning is easier at earlier ages for certain subjects, such as languages, the brain never loses its capacity to absorb new information. That's why curiosity plays such an important part in normal brain function during the later years.

9. *False.* When taken in moderation and combined with a healthy, nutritious diet, alcohol may actually enhance cognitive functioning.

10. *False.* The amount of sleep remains about the same. It is the quality of sleep that changes. As we age, we tend to spend more time in the lighter stages of sleep.

As we have discussed, it is likely that aging depends on both genetic and environmental variables: some people are simply more likely than others to live long lives. In addition, certain environments are so inherently unhealthy that people living in them will suffer reduced life expectancy no matter what life span their genes have programmed them for.

Added to these variables is the fact that each physiological system within the organism, each cell type within a tissue appears to age along its own trajectory. This is the explanation why people with, say, excellent cardiovascular systems may die of early-onset Alzheimer's disease. One organ may be aging in an optimal manner while another organ, due to genetics or environmental variables, may age at an accelerated pace. Thus aging must be recognized as dependent on cellular and organic "clocks" that at the moment are not well understood. One thing seems clear: all aging changes have a cellular basis, and ultimately, aging must be understood at the cellular level.

The brain is only one of several organs necessary for life, but it occupies a pivotal and privileged niche among bodily organs. Indeed, the health of the brain is the surest indicator of general body health. If the brain ceases to function normally, then the health of other body organs is unlikely to make much difference when it comes to the length, and certainly the quality, of survival. To learn this lesson one has only to encounter the pathetic sight of an otherwise healthy Alzheimer's patient who doesn't know the day or the time and can no longer recognize his relatives and friends.

And the brain differs from other bodily organs in another important aspect: no matter what advances may lie ahead in terms of increasing general longevity, not much can be done to reverse brain-cell loss once it occurs. That's because—with the exception of some olfactory neurons in the nose—neurons (brain cells) do not divide or make up for losses by spawning new neurons. Dementia is the result of the death of millions of these irreplaceable neurons. This is why it is so important to come up with treatments and cures for dementia.

But even though the brain cannot generate new neurons, it does have remarkable self-regenerating power. The story of a 78-year-old woman first described in 1990 in the *Archives of Neurology* illustrates this fact. Over the space of four years the woman showed a progressive deterioration in functioning. She lost memory, could not find her way around, and failed to recognize the people around her. A neurological investigation revealed she suffered from normal pressure hydrocephalus, an illness produced by an interference with the flow of cerebrospinal fluid within the ventricles of the brain. Treatment consisted of relieving the blockage by the placement of a shunt, which acted as a detour around the blockage.

Prior to the operation, neurologists and neurosurgeons carried out a battery of tests that revealed brain shrinkage along with significant reductions in regional brain-glucose utilization (34 to 49 percent of normal). In short, the woman's brain was sufficiently impaired that even her surgeons weren't confident an operation could accomplish much. Despite this, they proceeded.

Periodic testing carried out over the next two years showed a steady improvement in the woman's mental state. In parallel with the mental improvement, her brain volume increased, as did brain-glucose utilization in several brain regions.

Thus, even after four years of a seemingly permanent mental and brain deterioration, an 80-year-old woman experienced mental recovery and significant improvement in her brain volume and functioning. This example of brain plasticity and rejuvenation is particularly impressive since the woman was at such an advanced age when her operation took place. Until learning of her recovery, most neuroscientists believed that plasticity probably did not extend into the seventh and eighth decades. This operation demonstrates that all of the efforts at improving brain function suggested in this book are built upon a solid basis: the brain is capable of recovery and self-modification if provided the means.

"To add life to years, not just years to life" is the motto of the Gerontological Society. And the only way to guarantee that is to concentrate on the practical steps we can take now to guarantee that our brain remains at its best for as long as possible.

TWO

2

THE MENTAL

PERFORMANCE OF THE

MATURE BRAIN

hen speaking of mental performance, neuropsychologists use the word "cognition," a term for various aspects of thinking. Cognition includes memory, intelligence, attention, language ability, problem solving, and visual-and-spatial skills (usually referred to as visuospatial skills). Evaluations of cognitive performance in the mature brain take two forms.

Cross-sectional evaluations compare individuals of different ages. Thus, the memory performance of an 80-year-old might be compared with that of a 20-year-old. Presumably, whatever differences are found can be attributed to the effects of age. But there are problems with this approach. An 80-year-old didn't grow up with television and computers—his educational and social experiences differ in many areas that might have more to do with his cognition than the 60-year age difference.

The alternative approach involves *longitudinal sampling:* the same individual is studied over time. But this method of study also has problems. Adults rarely undergo neuropsychological testing unless something is already amiss. For this reason, tests carried out earlier in life when mental functioning was progressing satisfactorily are usually not available. In addition, there is a catch-22 involved in such testing: the people who are tested may represent only the successful members of a much larger group, many of whose members died long ago and are no longer around to be evaluated. Thus neither the cross-sectional nor the longitudinal approach is without problems—one of the reasons why disagreements exist about the nature and extent of cognitive changes in the elderly.

Tests of mental performance are actually measures of brain functioning. A continuum of activity exists, from the registration of simple stimuli (like light and sound) to highly abstract and symbolic activities (like physics or classical music). From the structural point of view, the brain can be thought of as an elaboration of the spinal cord. Ascending upward from the spinal cord, which carries sensory information to the brain and funnels motor commands through the arms and legs, the brain structures become increasingly more elaborated. Our thinking capacities, especially consciousness, depend upon the topmost portion of the brain, the *cerebral hemispheres.*

Each cerebral hemisphere is divided into lobes: frontal, temporal, parietal, and occipital. The *occipital lobe* mediates some aspects of vision; the *parietal lobe* manages sensation involving all the senses except smell; the *temporal lobe* is concerned with hearing and memory. Farthest to the front are the *frontal lobes,* which perform all of those mental processes that are unique to humans: our ability to form abstractions, foresee the likely consequences of our actions, and maintain a sense of personal continuity linking past, present, and future.

From the synthesis of brain activity in all of these areas (carried out within the brain's association areas) comes an internal model of the world that is the most evolved of any creature's. That's because our brain is capable of language and other symbolic representations. Only we humans have the capacity, thanks to our brain development, to form abstract concepts like time, to appreciate cause-and-effect relationships, and to anticipate and plan for the future. In order to do all this in an optimal fashion, the brain must be healthy. Brain disease at any age interferes with cognition. Diminished brain function in the early years results in the various forms of mental retardation; later in life, the corresponding defect is referred to as dementia.

Foremost among the measures of higher brain function is *intelligence.* Psychologists continue to argue about the nature of intelligence and how it is best measured. Most accept the distinction between *fluid* and *crystallized* intelligence.

All of the knowledge that you have learned over the years, use in your occupation, or refer to frequently makes up crystallized intelligence. Included here are vocabulary and general knowledge about how the world works. Like a crystal, it changes little over time.

Fluid intelligence, in contrast, involves gathering and using new information. Examples of fluid intelligence include responding to the uncertainties of the stock market and adapting to new personnel changes in an organization. Mental flexibility is another expression of fluid intelligence. The mature brain shows no change in crystallized intelligence, whereas fluid intelligence in an older person shows some decline. Brain scientists believe this difference reflects the brain's organization: crystallized intelligence involves brain structures that change very little over time, whereas the brain areas responsible for fluid intelligence are more subject to aging.

A second important measure of cognition is *attention.* If attention wanders or cannot be sustained, declines in performance can be expected in memory and language. Indeed, attentional difficulties underlie most complaints of "poor memory." Typically, the distracted or inattentive person doesn't pay attention to events going on around him or to the things people are telling him. As a result, the initial step in memory formation, *encoding,* never takes place. Without encoding, retrieval is impossible. You can't remember what you haven't learned in the first place.

With normal aging the mature brain experiences some decline in attentional power. This is thought to be due to a slight loss of brain cells from areas beneath the cortex. As previously mentioned, few neurons are lost from the cerebral cortex; in contrast to this small cortical-neuron loss, extensive loss occurs in special nerve-cell clusters *(nuclei)* located beneath the cortex. These special nuclei are in two-way communication with the cerebral cortex. By means of special chemical messengers *(neurotransmitters)*, they exert an alerting and

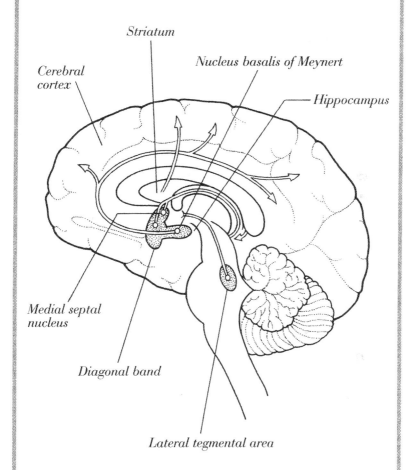

Striatum

Nucleus basalis of Meynert

Cerebral cortex

Hippocampus

Medial septal nucleus

Diagonal band

Lateral tegmental area

Acetylcholine (Cholinergic) Projection System in the Brain

energizing effect on the brain. Included among these subcortical nuclei are:

Nucleus basalis. Situated at the front and base of the brain, this nucleus releases the neurotransmitter *acetylcholine* in the cortex. Acetylcholine is considered a neuromodulator, a chemical that makes neurons more receptive to the action of other neurons—a key component of fluid intelligence. With normal aging this nucleus can be expected to lose as many as 50 percent of its neurons. And the level of activity of acetylcholine-producing neurons (cholinergic neurons) in this nucleus can affect information transfer and processing of the entire cerebral cortex. The result is a slight delay in coming up with specific information. The information isn't lost and can be retrieved; it just remains "on the tip of the tongue" for a little longer than would have happened years earlier. This delay is one of the reasons why, as a general rule, 70- and 80-year-olds aren't frequently encountered on quiz shows, where rapidity of response is just as important as the accuracy of the answer. It's not that the older person doesn't know the answer. Rather, he or she simply needs a bit more time in retrieving it.

With normal aging, as I said above, this nucleus can be expected to lose as many as 50 percent of its neurons. In fact, this particular nucleus loses cells with sufficient regularity as we age that it is used as the "gold standard" for determining brain age. In Alzheimer's disease, the numbers of cells that die off is far in excess of the expected 50 percent. The result of this loss is a tremendous falloff in fluid intelligence. (Crystallized intelligence suffers as well, but as a result of the loss of neurons from the cerebral cortex.)

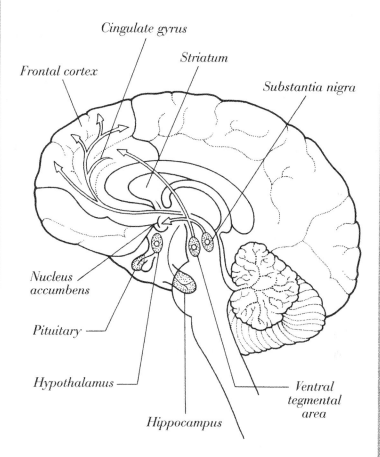

Cingulate gyrus

Frontal cortex

Striatum

Substantia nigra

Nucleus accumbens

Pituitary

Hypothalamus

Hippocampus

Ventral tegmental area

Dopamine Projection System in the Brain

Substantia nigra. This nucleus in the brain stem makes and distributes to the cortex and elsewhere the neurotransmitter *dopamine.* A 35 to 40 percent loss of cells occurs in normal aging. Since dopamine is involved in movement and emotions, the loss of cells in the substantia nigra can produce slowed movement and depression. Advanced cell loss results in Parkinson's disease.

Raphe nucleus. This nucleus produces the neurotransmitter *serotonin.* With normal aging as many as 35 to 40 percent of the cells may be lost without measurable result.

Hippocampus. While some parts of this nucleus suffer the loss, with aging, of as many as 35 percent of their neurons, those parts responsible for memory function (primarily early registration) lose no more than 5 percent. As a result, memory function is not much affected in normal aging. But in Alzheimer's and other diseases marked by serious memory impairment, the cell loss in these memory areas can be as high as three-quarters of the neurons.

Subcortical areas contain a small number of neurons that project upward toward the cortex in a fanlike distribution that affects the functioning of the entire brain. For this reason they are responsible for such basic and important processes as generating mental energy and maintaining alertness. Neuroscientist Paul Coleman refers to these nuclei as "the juice machines." Alterations in the projection systems emanating from these nuclei can increase or diminish underlying mental functions like arousal, attention, mood, motivation, and the level of anxiety. (Intelligence, largely a cerebral

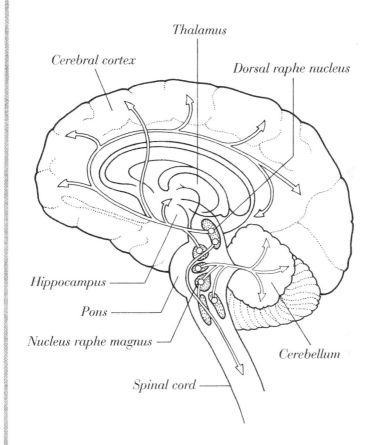

Thalamus

Cerebral cortex

Dorsal raphe nucleus

Hippocampus

Pons

Nucleus raphe magnus

Cerebellum

Spinal cord

Serotonin Projection System in the Brain

cortical function, remains unafffected, one of the reasons dementia is never the result of normal aging.)

With normal aging, attention becomes slightly more difficult to sustain and seems to require a greater amount of mental energy. Subjectively this takes the form of a subjective sense, with increasing age, of the need to put forward enhanced effort. This perfectly normal and easily correctable impediment to normal attention and concentration (corrective suggestions will be taken up in the "Pearls" section, in Chapter Nine) is the result of some slight falloff in the efficiency of the "juice machines."

A third component of cognition is *language*. The mature brain experiences no loss of facility in grammar, comprehension, or syntax (proper word placement and emphasis). Writers, for instance, do not suffer any loss in their language skills. Most older writers who have stopped writing have done so not from any loss in their language skills but simply because, like other people, they wish to retire. Tests of vocabulary typically show an increase rather than a decrease. Nor is an older person's store of knowledge usually affected (although it may momentarily appear to be affected since, as mentioned, immediate access to information may be slowed). Only one aspect of language performance is adversely affected with normal aging. An older person may experience failures in coming up with proper names. But this isn't abnormal or even restricted to the later years. Indeed the process starts in middle age. Anyone over 40 has experienced on occasion difficulty in coming up with a specific word despite the subjective feeling of having that word on the tip of the tongue. Most of us compensate for this failure by putting forth brief definitions of the elusive word or by substituting less precise terms, for example, referring to a dishwasher as a "household machine" or a "household appliance." Such momentary failures are perfectly normal in the mature brain and, unless part of a more pervasive pattern of deterioration, interfere very little with daily living. Fluency also declines with age. Speaking less often and more telegraphically, the older person may, depending on the circumstances, earn for herself a reputation for wisdom or taciturnity.

Memory is the mental capacity that changes the most in the mature brain. In addition, it is cause for the greatest concern among young and old alike. A more complete discussion of memory takes place in the next chapter. At this point, it's sufficient to note that memory changes do occur with aging, but they are limited to the rapid acquiring and retrieval of new information. Moreover, decreases in performance can be made up for by memory training and cuing procedures.

Visuospatial skills. The mature brain suffers some slight declines in depth perception, spatial localization, and the rapid identification of complex geometric shapes. But the impairment is usually mild. Often the performance impairment is the result of weakened eyesight, lack of education, and unfamiliarity or lack of interest in the testing procedures. Think about it for a moment: Why would an older person be all that interested in tests involving the recognition and manipulation of geometric shapes?

Problem solving. Traditionally tests for this rather vague category have consisted of laboratory tests or questionnaires. But the real test involves a person's ability to solve real-life problems—obviously a more elusive determination. In addition, problem solving cannot be divorced from attention, mental flexibility, and the talent for considering more than one way to solve a particular problem.

Concentration and focus. These mental functions are closely related to attention. With age most people are more vulnerable to distractions from both external and internal sources. They are most affected under conditions of high arousal, such as athletic competitions. In an athletic competition, the older person may, because of distraction, perform less well than a younger person (even allowing for differences in strength, speed, and endurance). For instance, a study of miniature golf involving younger and older players found that the older players often lost not because of their lack of skill (they had many more years of experience than their younger competitors) but because they were easily distracted. Most likely this heightened susceptibility to distractibility stems from subtle early impairments in the brain's frontal lobes, which are responsible for keeping subjects "in mind." Many neurologic textbooks attest to the importance of the frontal lobes as the brain structures responsible for the computerlike function of simultaneously keeping several different activities "on-line."

• • •

In contrast to the brains of all other creatures, the human brain possesses strikingly enlarged *frontal lobes.* Located at the extreme front of the two cerebral hemispheres and encompassing more than a third of the total cortical area of the brain, the frontal lobes, particularly the most anterior portions (the prefrontal lobes), are responsible for those features of our thoughts and behavior that distinguish us humans from other creatures. Included here are ambition and self-

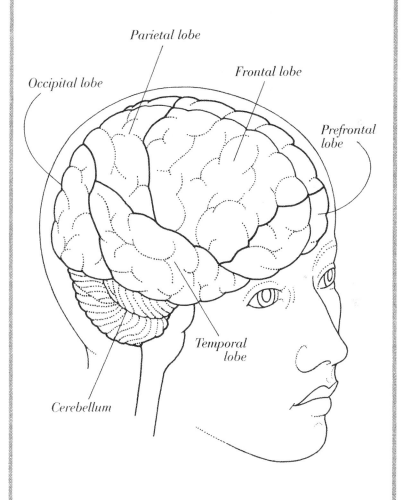

Parietal lobe

Frontal lobe

Occipital lobe

Prefrontal lobe

Temporal lobe

Cerebellum

motivated behavior; planning and anticipating the consequences of what we do; and the self-continuity that enables
us to maintain a sense of ourselves as changing and evolving
over time. Indeed, if you mention any human attribute that
is either missing or diminished in lower animals, it's likely
that that attribute is mediated by the frontal lobes.

With aging, a change in frontal-lobe efficiency takes place.
This principally involves the temporary storage and manipulation of information in the brain, a process psychologists
refer to as *working memory.* Working memory has a limited
capacity, lasts only a few seconds, and is most relevant for
planned, goal-directed behavior. For instance, if in the course
of a conversation in a crowded, noisy room your companion
tells you a telephone number, and a few moments later you
pick up a phone and dial that number without writing it
down, you are exercising working memory.

In an experiment testing working memory, young, middle-
aged, and older adults memorized the names of persons
standing next to them in a supermarket line while carrying
out exercises in mental arithmetic. The older participants in
the experiment performed less well than the younger ones.
Further, the results suggested a decrease with aging in the
capacity to attend to two or more tasks simultaneously.

In other experiments researchers monitored brain activation while volunteers performed two tasks either separately
or simultaneously. Activation of the prefrontal cortex occurred when the two tasks were performed together but not
when they were performed separately. Such findings support
the view that the prefrontal cortex is involved in working
memory (that is, each task must be kept in mind for later
resumption whenever attention is turned to the second task).

With increasing age, working memory becomes less efficient and reliable. An older person requires increased frontal activity in order to compensate for more rapid memory decay. Measurements of electrical activity accompanying exercises requiring working memory reveal increased frontal activation —almost as if the frontal lobes had to work harder to keep up. This finding is consistent with the decreased ability of the older person to maintain information on-line so that it can be used for comparison as part of working memory.

The decrease in the efficiency of working memory with age is an example of the last-in, first-out principle. Since the frontal, especially the prefrontal, cortex is the last portion of the brain to complete its development, this brain region seems to be affected very early during aging. The good news is that working memory can be enhanced through the use of stimulant drugs such as caffeine, amphetamines, and Ritalin. We will say more about this in the final section of the book, where specific "Pearls," bits of practical advice, for brain improvement in the later years are presented.

Cognitive decline with aging, when it occurs, stems from three major causes: disuse, disease, and aging itself. With increasing years, people often tend to use certain skills less. Not surprisingly, those skills suffer a decline. Physical illnesses too tend to increase with age; the various diseases affecting older people adversely affect cognition. Added to this are the normal neurological changes that occur with aging. Finally, there's individual variability: performance variability between any two people will increase as they age because of the cumulative effect of disuse, disease, and aging. As we will discuss in more detail later, the best defense against age-related cognitive decline is practice. Practice prevents disuse from occurring. In addition, practice can compensate for declines in brain performance by building up a reserve capacity to offset the brain changes. With practice, it's also possible to devise new ways of performing mental actions that will help maintain performance levels.

Summary of Normal Changes Accompanying Aging. Some decline in the mature brain can be measured in attention, certain aspects of language (mostly naming and fluency), some visuospatial skills, and memory. None of these categories *necessarily* declines, but all tend to do so unless special efforts are made to preserve them.

• • •

All of the above mental changes that accompany maturing were discovered by studying broad cross sections of the population: the educated along with those of limited education; those with preserved physical health along with those afflicted with one or more physical illnesses; those born into wealth and those raised in more humble circumstances.

For years specialists in the mature brain wondered what they might find if they tested a group of people who varied from each other by only a single factor, like education. What would be the differences between the typical aging person and someone who has maintained over a lifetime a high level of intellectual activity? Arthur P. Shimamura, a psychologist at the University of California, Berkeley, decided to find the answer to this fascinating question.

Shimamura and his colleagues tested seventy-two university professors and divided them into three groups: young professors, averaging 38.8 years of age (30–44 years); middle-aged professors, who averaged 52.2 years of age

(45–59 years); and senior professors, who averaged 64.7 years of age (60–71 years). Members of all three groups had about the same number of years of education (twenty-one to twenty-two years).

As a comparison group, Shimamura included in his study younger people (averaging 19.8 years) and older, less educated people (average 65.5 years). The older volunteers differed from the professors in the number of years of education (an average of sixteen years). "Thus the central focus of the study was to delineate the age effects among the professors," according to Shimamura.

Since the tests employed by Shimamura are illustrative of the methods employed by psychologists to measure the effects of aging, I'm going to describe them in some detail.

The first test measured raw reaction time. Subjects sat at a computer console and watched the screen for the appearance of a specific stimulus. In front of them was a response box consisting of a center key and four additional keys (above, below, to the left, and to the right of the center key). In the test of simple reaction time, the center key was pressed the instant a predetermined stimulus appeared on the screen. The stimulus could be either a word, usually "Go," or a traffic light symbol. In the two-choice test the word ("left" or "right") or symbol (arrow pointing left or right) cued the subject to push the left or right key. Finally, in the four-choice task, the word or arrow stimulus called for the key-press response to be up, down, left, or right.

The results? Older people took longer to respond than younger people, irrespective of education—the professors did no better than their less educated counterparts.

Shimamura next tested his subjects' ability to remember

and respond to arbitrary associations. This is something all of us try to do every day. We link a name with a phone number or an address. But frequently we forget the associations because there is no inherent logic or connection between a person's address and his name. Shimamura wondered whether aging would have any effect on learned associations.

In the face-face paired-association learning test, members of the three groups tried to memorize paired pictures of female and male faces ("Think of them as couples," Shimamura told his volunteers). After viewing six pairs of faces, the subjects viewed each of the women's faces again and tried to select the male face previously paired with it.

The name-name test consisted of ten name pairs (such as "Richard and Mary"). Following the presentation of all of the pairs, the subjects were again shown each of the men's names and asked to select from a display the corresponding woman's name.

As with the reaction-time experiments, the older professors did not perform noticeably better than the other older people when it came to learning paired associations.

To summarize Shimamura's findings so far, the two tests just described showed that education did not protect against slowing. In Shimamura's words, "intelligence and mental activity during adulthood cannot prevent problems in memory for arbitrary associations, such as associating names to faces or phone numbers to colleagues."

Shimamura's findings on a slowed speed of response with increasing age agrees with a long history of research on aging. Psychologists have known for over a century that as a person ages, he or she processes information more slowly. This results from a slowing of information processing within the brain itself, rather than from delays in the conduction of impulses along the nerve fibers traveling back and forth between the brain and the outside world. This age-associated slowing of information processing within the brain explains why many older people show a falloff in performance on tests measuring rapidity of response involving memory, reasoning, knowledge, and fluency.

Neuroscientists believe that speed of processing should be viewed as one element, albeit an important one, of the architecture of the brain as it evolves across the entire life span. As a person ages, a balance must be achieved between the time required to maximize speed and the care needed to maintain a high level of accuracy. In the older person especially, the likelihood of errors increases as the time frame for the responses progressively shortens. As mentioned earlier, this is one of the reasons you don't see many older people excelling on *Jeopardy* or other game shows where knowledge must be combined with a rapid response.

At this point Shimamura dug a little deeper and in the process discovered some very good news about aging and the brain.

•   •   •

Shimamura showed his subjects a computer-generated display of sixteen visual patterns on a single television screen. The set of patterns was then repeated in random fashion—the sixteen patterns remained the same but were arranged differently. With each repetition, the subjects were asked to point to a different pattern—one they had not chosen before. To do this, of course, they had to remember what they had already pointed to from one presentation to another, over

sixteen trials. Shimamura increased the difficulty by repeating the test using the same patterns in a second trial. It
was this second trial that distinguished the university professors from other subjects of the same age.

As a general rule, the mature brain performs as well as a
younger brain on single-trial testing. But when the visual
patterns are shown for a second time, most older people have
difficulty discriminating between responses from the first trial
and those made during the second. The psychological term
for this inability to ignore or inhibit irrelevant information
from an earlier trial is *proactive interference.* It is thought to
result from failures in frontal-lobe functioning (patients with
disease of the frontal lobes characteristically show dramatic
proactive interference). This falloff in performance with repeated testing in the mature brain correlates with neuronal
loss resulting from normal aging, a process that affects the
frontal lobes more than any other cortical region.

But the professors in Shimamura's study tested differently.
In contrast to the other people in the control group, none of
the older university professors showed proactive interference.
They did as well in the second trial as in the first.

In the final test administered by Shimamura, the subjects
listened to three tape-recorded stories. The first was a fic-
tional account of the robbery of a woman identified as Anna
Thompson. The story contained twenty-five key phrases or
facts, and each subject was asked to recall as many as possible
of these twenty-five items. The second story involved scientific facts about the elements that make up the earth's atmo-
sphere. The third story was about tribal cultures in
prehistoric times. Again in both the second and third stories,
subjects were rated according to how many of twenty-five key

phrases they could successfully recall. In this test too the older professors performed as well as younger professors and significantly better than the group comprising less educated seniors.

Shimamura's research findings show that as it matures into the later years of life, the brain demonstrates a slowed reaction time and is slightly less facile in learning and associating unrelated items. Education appears to exert little effect here. As a practical consequence, "intelligence and mental activity during adulthood cannot prevent problems in memory for arbitrary associations, such as associating names to faces or phone numbers to colleagues," according to Shimamura. (In the last portion of the book we will suggest ways of compensating for this deficiency.)

But the good news is that the mature brain performs as well as a younger one in tasks requiring planning, organization, and the manipulation of information. Thus the mature brain is not simply a defective version of a younger one. With advancing years, the brain redesigns itself to compensate for decreases in reaction time and general slowing. Rapid retrieval of information becomes less important than the application of the information.

A good education and an occupation requiring a high degree of mental activity leads later in life to maintained and even enhanced mental performance in some areas. The experience of the older person enables him or her to put information into a context or a specialized-knowledge-based framework.

As a general rule, younger people may come up with scattered bits of information faster but lack the experience needed to put that information into context and draw the most appropriate conclusions from it. The older lawyer, the more experienced surgeon, the elderly statesman, the senior corporate manager—these individuals excel not on the basis of rapid information retrieval but because of their ability to draw upon their accumulated knowledge to reach important conclusions and make key decisions.

Shimamura believes the professors perform better than their less educated peers because of efficient planning, organization, and retrieval strategies:

Perhaps professors are facile in their use of such strategies to access knowledge as a result of daily activities that require the integration of new knowledge into existing knowledge representations. Their facility with these strategies may compensate for slowdowns in the mechanics of cognitive function, such as slowdowns caused by cognitive slowing.

Shimamura's results would not have come as a surprise to the Latin orator and philosopher Cicero. Writing around 44 B.C. in his *On Old Age*, Cicero enjoined those who would wish a healthily functioning mind to care not only for their body: "Much more care is due to the mind and soul; for they, too, like lamps, grow dim with time, unless we keep them supplied with oil. Moreover, exercise causes the body to become heavy with fatigue, but intellectual activity gives buoyancy to the mind."

Although no one knows for certain why education provides protection against some important age-related changes in mental functioning, the explanation favored by most neuropsychologists is that mental activity plays a compensatory role for normal age-associated losses in these areas. And that explanation seems to make a good deal of sense. If increased mental activity reduced all age-associated brain changes, then slowed reaction times and other signs of age-related declines would not have been found in the professors. The fact that such signs of brain aging were found among the professors suggests that age-related changes occur to some extent in everyone, including the better educated and mentally vigorous like the professors, but that education and mental activity can protect against losses in other important areas of brain function.

Another study comparing elderly eminent professors with blue-collar workers of comparable age confirmed Shimamura's findings. The researchers found that

Initial ability determined the level of intellectual performance such that elderly academics maintained their initial advantage over the elderly blue-collar workers. However, the rate of change on tests of memory and intelligence did not differ for the high- and low-ability groups. Thus high ability is not associated with a slower rate of cognitive decline.

In short, intelligence and memory decline at a similar rate for well-educated and less well educated people, with the better educated declining to a lesser absolute level than those individuals who were at a lower point educationally to start with. This means that older people in the workplace, particularly professionals with a lengthy experience in an area of specialization, can compete successfully with their younger counterparts.

"It does not necessarily follow that increased age is a handicap in most occupational situations because experience may play an important role in moderating the influence of age on work performance," according to T. A. Salthouse of the School of Psychology at the Institute of Technology in Atlanta. Salthouse's conclusion has practical implications in regard to work, career, and age of retirement.

Zaven Khachaturian, formerly of the Neuroscience and Neuropsychology of Aging Program of the National Institute of Aging, now an international lecturer and consultant on aging, says:

We as a society are terribly remiss about how we treat our older citizens. It's wrong to place all older people in the same category. One person loses his functional capacity before age 65 and should rightly be retired. But what about the Picassos who can function creatively into their 90s?

In the business world a company should treat its employees at least as well as it treats its physical plant. As things stand now there is a huge disparity between how a company decides on "retiring" its industrial plant and how that same company goes about handling their older employees. No company would close down a plant that remains productive. So why close people down who still demonstrate capacities for innovation and creativity? Society should stop making decisions about a person's usefulness on the basis of age and start making those decisions on functionality.

3

MEMORIES

ARE MADE OF THIS

*A*s we get older, we tend to worry more about our memory failures than about any other aspect of our mental functioning. This is because of the well-publicized association between memory loss and Alzheimer's disease. As a result, even a moment's hesitation in coming up with a name is feared to be a harbinger of serious and irreversible brain disease.

One of the surest ways of getting some perspective on memory performance in the later years is to consider the basic nature of memory: How do we form memories? Do we remember things as they were, so that in order to re-create the original experience, we need only make a selection from a kind of mental videotape library and put it into the equivalent of our brain's VCR? Or is memory a dynamic constructive process that changes over time and can never be "frozen"?

Experiments and everyday observations provide great variations in what different people recall about the same event. In an experiment carried out by psychologist Ulrich Neisser, individuals chosen at random were asked to recall where they were and what they were doing when they first heard of the *Challenger* disaster. They were asked the same question a year later. Their memories had changed. Moreover, when confronted with the discrepancy, the subjects either denied making the earlier assertions or expressed confidence that their second recollection was the correct one. There was also a poor correlation between a subject's confidence in the accuracy of his recollection and just how accurate he actually was. Such discrepancies should not be surprising, of course, to anyone who lives with another person: details about when and how things happened routinely form the basis for family quibbles.

Nor does the starkness or vividness of the original experience guarantee accurate recall. One would expect that something like the *Challenger* disaster would stand out particularly clearly and therefore be remembered with vividness and accuracy. Yet it wasn't.

Clearly the videotape analogy for memory is wrong. Memory storage within the brain isn't at all like a videotape. It's an active, dynamic process. Indeed, the past is almost as fluid, uncertain, and dynamic as the future.

For instance, if we are depressed we tend to remember all of the sad and discouraging things that happened to us over the years. When we're in such a frame of mind, even our past successes seem paltry and insignificant. Later, when our depression lifts and the world doesn't seem so bad, we start remembering once again all of the good things that happened to us.

Memory can be thought of as having two compartments: a *temporary* store, holding information for a brief period of time, and a *permanent* store, where the conscious ability to remember previous events is located. (This latter is called declarative memory.) As a first step, material to be remembered must be registered and retained. In general this aspect of memory changes little with age. A person's ability to hear and repeat back a telephone number or a brief grocery list is as good at 70 as it was at 20 or 30. (Distraction or inattention will worsen the performance, of course.) The primary problem with memory loss due to aging involves the transfer of information from the temporary store to the permanent store.

All types of memory are dependent on the temporal lobes, particularly their anterior and medial poles, along with their connections to two structures comprising the limbic system, the hippocampus and the amygdala. The frontal lobes are also called upon since they contribute to the attention and concentration required for learning.

Immediate memory requires a normally functioning medial temporal lobe with intact connections to the adjoining hippocampus and amygdala. Problems anywhere in this circuit result in a severe memory disturbance, the so-called amnestic syndrome. The most famous instance of this was H.M. (only his initials are used in order to preserve his anonymity).

In 1953, while in his early 20s, H.M. was operated on to cure him of a severe seizure disorder. The neurosurgeon excised the anterior medial part of both temporal lobes, the amygdala complex, and the anterior two-thirds of the hippocampus. Now in his 60s, H.M. cannot form new memories and forgets what he has learned within moments. If you are introduced to him and walk out of the room and return after

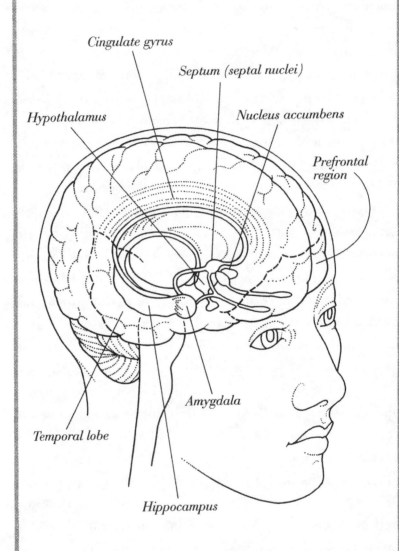

Cingulate gyrus

Septum (septal nuclei)

Nucleus accumbens

Hypothalamus

Prefrontal region

Temporal lobe

Amygdala

Hippocampus

a few minutes, he will not remember having previously met you. Neuroscientists now know that the hippocampus and the amygdala are critical in the initial memory step of encoding. What cannot be encoded cannot later be recalled. We speak here of a retrieval rather than a recognition problem: H.M. is often able to select the correct answer from a multiple-choice test but cannot come up with the same information if asked for it directly. This is because recognition memory is more distributed throughout the brain and is dependent on different structures, which bypass conscious awareness. We all know more than we can say and more than we think we do.

Prior to examining memory failures that result from brain dysfunction, it's useful to consider normal forgetting. The most common causes of forgetfulness are distraction and inattention. Both interfere with the initial encoding of the information; this, in turn, precludes storage and subsequent retrieval. Distraction and overwork, at any age, interfere with encoding and frequently result in the distracted person seeking medical or psychological help for "memory" difficulties stemming from these encoding problems. Actually, the memory is quite normal under such circumstances. What's needed is not medical treatments or psychological help but greater attention to the material to be learned.

• • •

While memory failure seems like a fairly objective matter that most people can agree about, judgments may unwittingly involve stereotypes. Surely you've often heard repeated this stereotype: older people remember the distant past very well; it's the recent past that they have problems with.

Actually, memory performance with age declines equally

from the distant past right up to the present. A study carried out in Japan involving subjects ranging in age from 40 to 79 revealed that with increasing age, memory performance declines equivalently for events and information across the decades. In short, aging influences very old memories as well as recent memories. Thus the image of an older person's memory as dwelling in the past but incapable of registering things in the present is a cruel distortion. In fact, it is perfectly normal for memory to decrease slightly with aging, but this decrement does not selectively involve more recent periods in favor of earlier years.

A second stereotype concerns the interpretation put upon memory failures. In practice, the interpretation that we put upon another person's memory failures depends very much upon distorting preconceptions that most of us are not aware of. In a fascinating experiment carried out at the Department of Psychology at Florida International University, experimental subjects were asked in a questionnaire their opinion about the relative influence of a person's age and attractiveness upon judgments about that person's memory failures. The subjects overwhelmingly selected age as the more important contributor to judgments about memory failure, with attractiveness playing little or no role.

In the second part of the experiment, the subjects listened to recordings of young and elderly women reading aloud descriptions of short-term, long-term, and very long-term memory failures. While listening to these recordings, the subjects had at hand pictures of attractive or unattractive young or elderly women. As each recording was played, the test subjects were directed to the picture that supposedly corresponded to the person heard on the recording (actually, there was no correspondence between recording and picture).

After hearing the recordings and seeing the pictures, the subjects were asked to judge whether the memory failures described in the recordings were the result of mental difficulties and whether a medical and/or psychological evaluation was indicated. The result? Memory failures in young, attractive women were attributed to lack of effort or attention, while unattractive people of any age were more likely to be considered in need of evaluation and treatment. Unattractive elderly women were deemed the most impaired and the most in need of some type of medical treatment. Thus, despite earlier declarations to the contrary by the participants in the study, attractiveness rather than age actually served as the determining factor in this experiment that closely simulated how judgments are made about memory failure in real life.

This study is important to remember when considering the seriousness of a memory impairment in an older person. Mental abilities like memory, while objectively measurable, are somewhat uncertain and subjective. Failures of recall may have nothing to do with memory. For one thing, the mature brain is more vulnerable than the younger brain to the adverse effects of anxiety, depression, and other negative emotional states. All of these interfere with recall. This means that even in otherwise healthy elderly individuals with healthy brains, negative emotional states can impair memory.

Other assumptions about memory in an older person may sometimes be true, sometimes only a stereotype. For instance, the older a person is, the less likely he or she will benefit from a mix of verbal and pictorial instructions. A study comparing comprehension and memory for drug information in younger and older people revealed that the instructions on medicine bottles are comprehended better in younger people if the instructions comprise a mix of verbal instructions and pictorial representations. Older people reacted just the opposite and did best with well-organized verbal prescriptions. Such a finding shouldn't be surprising. Younger people are more accustomed to dealing with mixed verbal and pictorial representations, such as are encountered in video displays and many of the most popular word processing programs.

• • •

Before exploring the nature and causes of memory failure, it's useful to consider that normal memory in the healthy older person has unique characteristics. Psychologists have found that an older person has a hard time with memorization or coming up with specific answers to direct questions of fact. While my informal investigations tend to confirm this observation, I can also recall my grandfather reciting the endlessly fascinating (at least to me at age 12) stanzas of "Casey at the Bat."

Clearly it's not so much recall of old memories as committing new material to permanent memory that provides the greatest challenge for the old. Indeed, the older we are, the harder it is to memorize anything (perhaps one of the reasons that acting is rarely taken up as a retirement hobby). Proper names are particularly difficult to recall. One factor contributing to this difficulty is that names are arbitrary labels: there is no reason why every Robert in the world couldn't just as well be named Ronald.

Another problem arises from the fact that so many people share the same names. The older you are and the more people you know, the more likely it is that you will become acquainted with individuals sharing the same first name. Also, people haven't just one name but two, and the first and last names have to be learned as an associated pair. But remembering a person's first name carries no guarantees you will remember his or her last name, and vice versa.

Of course, none of these challenges are unique to the mature brain. Recalling proper names is a task we are faced with from the moment we first begin to talk. But the older we get, the more difficult it becomes for our more mature brain to accomplish this name-recollection task successfully. This decrease in memory power can be compensated for by any method that increases distinctiveness. For instance, "Reginald Fairborne" (tall, blond-haired, rather pompous young man of 32) is an easier name to remember than "John Smith" (40ish nondescript minor government employee). That's because a name like "Fairborne" is unusual and carries with it the implication of a person born into conditions of wealth and privilege. Complementing this association is the relative pomposity conveyed by the rare and stuffily formalistic-sounding "Reginald." In short, it's easy to form a memorable mental picture of Reginald Fairborne, while John Smith lacks distinctiveness and fades into anonymity.

Books and exercises to improve memory are all based on increasing the distinctiveness of proper nouns, usually through the creation of dramatic and even bizarre mental images. The most famous of these and the one that forms the basis for the various mnemonic systems is the *method of place*, allegedly devised by the Greek poet Simonides.

As Cicero describes Simonides' method, "We ought to set up images of a kind that can adhere longest in the memory.

And we shall do so . . . if we set up images that are not many or vague but active." The images should be novel or marvelous, since "ordinary things easily slip from the memory while the striking and the novel stay longer in the mind."

Specialized memory techniques based on Simonides' method live on two thousand years later in many memory courses and popular memory-building books, such as *The Memory Book* by Harry Lorrayne and Jerry Lucas.

But while an older person may suffer some decline in performance when asked to remember a specific name or proper noun, he will usually have little difficulty choosing the correct one if offered several alternatives. That explains why the elderly man or woman having trouble answering a question will have no trouble in recognizing and choosing the correct response if it is encountered as part of a multiple-choice test.

Our understanding of memory and the importance of the temporal lobe and its connections to memory formation in the mature brain has been furthered immeasurably by PET (positron-emission tomography) scan studies. PET images show that the young and the old remember the same way, via activation of the hippocampal regions on both sides of the brain. This supports the view that, at least in the early stages of memory encoding, the same processing is taking place at all ages. But the PET studies also show important age-related differences.

One PET study of facial recognition provides an excellent example of how the mature brain goes about remembering. The study focuses on the processes that take place when a person observes and later recognizes human faces.

In an experiment at the National Institute on Aging, neuroscientists observed groups of young (23–27) and older (64–76) people performing three tasks while undergoing PET scans. In the first experiment, the subjects memorized a set of thirty-two unfamiliar faces (encoding). They then matched two different sets of faces (perception). Finally, they were challenged to recognize faces from the original set that were mixed in with "distractor" faces (recognition). While no differences appeared in the two groups in terms of reaction time, accuracy was lower for the older people in the recognition trial. The investigators found that the lowered recognition abilities were reflected in PET scan readouts, which showed a lowered hippocampal activation during encoding.

The researchers speculate that the age-related impairment of recognition memory may be the result of the disturbance in hippocampal activation shown in the PET scans during encoding. This makes sense, since the hippocampus is thought to participate in storage of new memories by collecting and synthesizing impulses from various brain regions. In fact, neuroscientists believe hippocampal activation is an index of the accuracy of recollection. If the initial encoding process is interfered with, then later recognition cannot take place—you can't recall a telephone number told to you while you were daydreaming or otherwise distracted. In an older person, despite his or her best efforts, the brain may fail to engage the circuit necessary for making a memory trace that can later be accessed.

Other compensatory changes may occur in the mature brain. If the recollection is incorrect or uncertain, the frontal regions of the brain are activated in an effort to force recall. It is this effortful recall that best distinguishes the young and the mature brain.

In the younger person both frontal areas are activated, while in the older person the encoding in the left hemisphere seems to be reduced in comparison to a comparably normal right-frontal activation. This difference in the activation in the frontal areas is thought to contribute to the declining memory performance of the older compared to the younger person. It's as if the older brain has difficulty activating the parts of the brain associated with effort. This reduction, along with the reduced hippocampal activity, accounts for the re-ductions in PET scan activity in these areas.

According to Cheryl Grady, leader of the team of research-ers who carried out the NIA study, "The degree of memory impairment with aging varies, but we find that older people have a harder time storing new memories than do younger people."

Alterations in vision and hearing may also play a role in memory formation. An older person with a mild hearing im-pairment memorizes a word list better after reading it than after hearing it read or repeating it aloud himself. This is because the extra effort required to recognize spoken words in the presence of a hearing impairment prevents the needed rehearsal and encoding of the material to be memorized. Ob-viously something cannot be committed to memory until it has been clearly heard.

But even in the presence of normal aging and normal age-adjusted decreases in hearing ability, memory formation can be interfered with. For instance, speech and hearing experts have discovered that speech perception can deteriorate as a result of two processes. The biggest contribution comes from the (sometimes slight but always measurable) progressive high-frequency hearing loss that normally accompanies aging. A smaller contribution comes from a general slowing of performance. Both of these components may affect memory.

As a practical consequence, a mature person with good eyesight is likely to learn more efficiently from reading rather than listening to the same material. This general rule of thumb is routinely ignored in most adult education courses, where the participants spend the class time listening to lecturers and only later reading supplementary material. While learning can be enhanced by both listening and reading, it's useful to remember that in some instances memory problems may be the result of defects in hearing.

• • •

The *key points* about memory are these: Everyone above 50 experiences some problem with memory. The distinguishing question, therefore, is not "Am I having a problem with memory?" but rather, "Am I having enough of a problem that it is seriously interfering with my life?" Or, to put it differently, "Is anyone else complaining about my memory?" In instances of severe memory loss, the signs will be observed by others.

Recently doctors from the Free University in Amsterdam developed a four-question test that reliably distinguished normal forgetfulness from forgetfulness requiring more extensive evaluation:

- Do you have memory complaints?
- Do you tend to forget the names of relatives and close friends more often than you did in the past?
- Do you often misplace items or forget where you left them?
- Have you lost your way on familiar streets?

Anyone who answered yes to two of the four questions was more likely than other people to score poorly on more elaborate and thorough memory tests. The four-question test, however, did not determine whether people who did poorly were more likely than others to go on to develop more severe memory problems, such as occur in Alzheimer's disease.

In the absence of a "gold standard" for memory, a useful dictum suggests itself: only when one's memory ability is severely deficient compared with those of others of the same age, interests, and education should there be cause for concern. For memory loss to be a sign of disease, a falloff from the usual performance should be apparent not only to the person involved but to those around him or her.

•   •   •

These findings are cause for a great deal of optimism and hope. All of us undergo periodic fears that our memory is seriously failing and we are perhaps coming down with Alzheimer's disease. But at such times remember that, although alterations in memory exist in normal aging and Alzheimer's disease, the nature and degree of the memory changes differ dramatically. For instance, if you were even in the early stages of Alzheimer's, it would be extremely unlikely that you would be able to remember having read up to this point in the book. That's because your brain would be incapable of registering and retaining enough information from the early chapters for you to process and make sense out of what you're reading now.

While it's true that some delayed recall occurs as we age, it is not limited to older people. It starts at least by age 50, does not interfere with occupational success, and can be compensated for by increased effort and novel learning strategies.

Much of the memory difficulty we experience later in life results from the increased effort required by the mature brain to achieve the initial encoding. But once the memory is formed, no significant age differences exist in terms of how successfully the information can be utilized. This leads to what I consider a pivotal insight: when we compare the young and the mature brain, differences in memory performance, and performance in general, usually involve energy and effort. The mature brain can achieve as much as its younger counterpart, but it has to work harder at it. The issue is not a loss of brain cells or a decrease in nerve-cell networks. What's different is the energizing factor, and this can be compensated for by several strategies. Sheer force of determination works for many successful people. For those who cannot generate increased energy and enthusiasm, energizing drugs are available by prescription to aid in the process. In Chapter Eight we will take up these chemical memory aids.

The older person is more vulnerable than a younger one to the adverse effects of depression and other negative emotional states. Therefore, even in normal elderly individuals, anxiety and depression may interfere with memory function. Psychiatrists and neurologists observe this firsthand. An elderly person may have difficulty remembering routine matters on an initial consultation, and yet weeks later, after the successful treatment of depression, the person's memory is perfectly fine.

Specific activities provide protection from memory impairments. Psychologists from Scripps College in Claremont, California, compared fifty bridge players and fifty nonplayers between the ages of 55 and 91. All the participants took part in tests of memory, reasoning, reaction time, and vocabulary.

Results showed that the bridge players outperformed non-players in measures of working memory and reasoning, but not vocabulary and reaction time. The results were consistent with the fact that bridge provides players with specific experiences in the exercise of memory and reasoning; vocabulary and speed of response play only subsidiary roles in bridge.

Psychologists at Northern Illinois University studied another skill-related game, bingo. Bingo, in contrast to bridge, does reward speed of response along with memory proficiency. The research project included players ranging in age from 19 to 74. No age-related differences were found on memory measures—"suggesting that Bingo playing experience may have positive benefits for many older players. The oldest and most experienced players did not differ from the younger equally experienced players on the cognitive and skill-based tasks," the researchers concluded.

Both bridge and bingo are games of skill, and both involve a good bit of socialization and interaction. This interactive component no doubt also plays a role in sustaining more efficient memory and cognitive performance. Both the bridge and the bingo studies emphasize how leisure recreational activities can provide sufficient intellectual stimulation to maintain healthy memory capacities comparable in efficiency to those of people decades younger in age. The same can be said for memory involving perceptual motor tasks.

Age makes no difference, for instance, when it comes to learning and remembering how to operate the mouse control on a computer screen. Tests show that while older people often take longer to develop skills in mouse control, they maintain their skills over time at the same level as people many years their junior.

According to psychologist Marilyn Albert, improvement in memory performance in the later years involves six specific steps.

First, *pay attention* to what it is you are trying to learn. Most so-called memory failures result from lack of concentration. "The true art of memory is the art of attention," wrote Samuel Johnson.

Second, and related to the first step, *make learning conscious and deliberate.* Keep in the forefront of your mind what you are trying to learn and why it is important to you.

Third, *make what you need to remember more interesting, more connected, and more meaningful.* The various memory systems and aids stress the formation of vivid mental images that incorporate the material you are trying to learn.

Fourth, *pace your memorization efforts.* With increasing age it takes longer than it did a few years earlier to learn the same amount of material. Once learned, however, the material is retained equally well across the age span.

Fifth, *use memory aids.* This can involve various memory systems or perhaps nothing more elaborate than a small notebook containing lists and schedules.

Sixth, *be consistent in how you use your memory aids.* If you opt for a notebook, then use it daily. Make it a regular companion that you take with you everywhere. Such use of a memory aid, incidentally, does not have to mean you become dependent on it to the exclusion of your unaided memory. Before you look at the list you composed earlier in the day, see how many of the items you can spontaneously recall from memory. Only then check your list.

Cicero anticipated the modern concept of memory functioning in maturity:

But, it is alleged, the memory is impaired. Of course, if you do not exercise it, or also if you are by nature somewhat dull. . . . I certainly never heard of any old man forgetting where he had hidden his money! The aged remember everything that interests them, their appointments to appear in court, and who are their creditors and who their debtors. And how is it with aged lawyers, pontiffs, augurs, and philosophers? What a multitude of things they remember! Old men retain their mental faculties, provided their interest and application continue; and this is true, not only of men in exalted public station, but likewise of those in the quiet of private life.

4

ROBUST AGING

'm writing this on a morning when two longevity-related stories are featured in the news. Over breakfast I'm reading an article in the *Washington Post* about why scientists at the National Institutes of Health have concluded that beta-carotene, a heavily promoted vitamin supplement, not only does nothing to protect against heart disease or cancer but fails to increase longevity (the researchers even found some evidence that the food supplement actually increased cancer rates among heavy smokers). In addition to the finding on beta-carotene, experts quoted in the article recommended healthful diets low in fat, vitamin ingestion, exercise, and regular physical checkups.

Meanwhile, on one of the early-morning shows, George Burns is being interviewed on the occasion of his 100th birthday (he died a few months afterward). Burns is smoking a cigar, and the interviewer asks him, "How many cigars do you smoke a day?"

"Somewhere between ten and fifteen," Burns responds.

"And how about drinking? You have a few drinks every day too, I'm told. How many of those?"

Burns responds, "Between three and five martinis every day."

The interviewer looks incredulous. "And what does your doctor say about that?"

Burns takes a drag on his cigar and responds without missing a beat, "My doctor is dead."

How does one balance these two very different approaches toward health and longevity? While most scientists would not suggest ten to fifteen cigars and three to five martinis a day as a sensible approach to living a long life and retaining one's mental agility, George Burns was living proof that a person can do this. How did Burns and others like him manage to live so long and so successfully? This question is particularly intriguing to us because we all would like to live to be 100, assuming good health, friends, and independence. But how do we go about increasing the odds of reaching that magic age of 100? One important factor stands in the way of a universal formula for a long life.

Each of us inherits a certain genetic disposition for a host of health-related factors, including longevity. A man like Burns was living proof that there are no absolute rules or limitations when it comes to living a long life with fully retained mental capacity. But what worked so well for him is likely to get many of the rest of us into a great deal of trouble. Rather than indulging in cigars and martinis, most of us remain more abstemious, generally preferring to place our bets on some currently available or yet to be discovered chemical fix.

Beta-carotene is only the latest in a series of "magic capsules" that on closer examination have turned out not to be magic after all. This doesn't mean, however, that we should embrace a nihilistic, "I guess I can't do anything" approach.

Scientific research confirms that we can do a lot to increase the likelihood of a normally functioning brain in our later years. But we should approach these activities with the realization that trying to predict the course of anything in life, especially something as complicated as mental and emotional health during the last quarter of the life span, is fraught with difficulty and uncertainty. The most we can hope to do is turn the odds in our favor.

All of us will age along one of only three general projectories: *successful aging, usual aging,* or *diseased aging* (this last means aging accompanied by one or more aging-associated illnesses).

All of us know older individuals who have aged very little over the years. These people are as intelligent and mentally competent now as they were twenty years ago. Unfortunately, neuroscientists are not able at the moment to explain how such people retain their mental competence. Nor are they able to tell us measures that we can take that will *guarantee* a similar course for us.

• • •

The mental decline that may accompany aging stems from three sources: normal aging (as previously described), disuse, and disease.

All of us tend to use certain skills or abilities less as we grow older. As a result, these skills decline. After a few months' absence from the bridge table, we can no longer engage in challenging and competitive play. After an absence of a few years, the bridge skills encoded into our brain over many years of active play have atrophied to the point that to resume serious bridge again we have to take lessons.

Physical illnesses also increase with age, and these often affect thinking and emotional health. Some diseases, such as those attacking the heart and lungs, directly diminish blood flow to the brain and thus impair mental alertness and cognitive functioning. The pain of other illnesses distracts the mind; sometimes, as with cancer pain, the mind is completely preoccupied.

Finally, there are degenerative illnesses like dementia and strokes, major and minor, that injure or destroy nerve cells directly and thus diminish the brain's functional capacity. But in the absence of these destructive processes (all of which can be favorably influenced, as we will discuss in a later chapter), there is no reason we cannot inch closer to that goal of reaching 100 years of age. What distinguishes successful centenarians?

In order to answer that question, a large number of geriatric and gerontology centers throughout Italy joined to form the Italian Multicenter Study on Centenarians. Although the census is still ongoing, as of December 31,1993, six thousand centenarians had been located and interviewed.

Certain factors, it turns out, correlate with *centenarianism*, or extreme longevity. Centenarians tended to have followed a balanced diet based on natural foods. During their productive years they were hard workers with a greater than average degree of self-sufficiency and independence. Though not wealthy, they tended to live for long periods in familiar and comfortable surroundings. They maintained normal choles-terol and blood-sugar levels. (The investigators apparently neglected to inquire about cigars and martinis.)

According to another survey, in Shanghai, centenarians en-joyed good medical care and the "benevolent attention" of family members. One centenarian, a 115-year-old woman, the oldest individual in China as of September 1994, was examined and interviewed by physicians for the purpose of learning "the secrets of her social and medical profile." She attributed her long life to heredity, well-balanced meals, suf-ficient sleep and exercise, and no smoking or drinking. Her doctor agreed with all of that and added to the list the excel-lent medical care he and other doctors had provided their patient during the many episodes of pneumonia and urinary-tract infection she had contracted since age 109.

A direct-mail questionnaire of 398 centenarians living in Japan in 1992 turned up two intriguing findings. Those with higher levels of education were more frequent among cente-narians (27 percent of men and 8 percent of women) than among the general population (less than 2 percent). Second, the average age at death of the parents of the centenarians was significantly higher than the average life span in the last decade of the nineteenth century—a suggestion that longevity was to some extent inherited in these centenarians' families.

Whatever the contributing factors to longevity, one thing seems certain: a person can survive into extreme old age with his or her brain functioning at a high level. A historical and postmortem study of twenty institutionalized centenarians who died at Charles Foix Hospital, Ivry/Seine, France, turned up "a low proportion of demented patients." This is even more remarkable in light of the fact that the sample omitted those centenarians least likely to be impaired: those living alone or with family, rather than in institutions.

"We have encountered optimistic, wise individuals very engaged in living," concluded a report on centenarians made by the National Institute of Mental Health–funded Georgia Centenarian Study. "Our research team continues to be amazed at the quality of life and the individual differences amongst centenarians."

The point here is that the oldest old display the same variation in brain performance and general health that occurs at every other point in the life span. "There is no age limit for the maintenance of inter-individual differences in aging. And there is no general age limit for the maintenance of independence and social competence," concluded U. Lehr of the Institut für Gerontologie, Universität Heidelberg, after analyzing international studies on centenarians.

The findings on centenarians should be put into context. By the year 2030 the United States will be the home of twenty-five thousand people over 100 years of age. The U.S. population then will contain as many 65-year-olds as 21-year-olds. The positive finding on centenarians should also serve as a correction to the frequently voiced opinion that life after middle age involves an inevitable downward spiral toward disability. These studies of individuals who have lived to be 100 suggest that the brain possesses the potential for superior performance beyond the ninth decade.

• • •

Although most of us have never met a centenarian, we all have certain stereotypes about an older person. One of the stereotypes is that an older person is a wise person. We expect that a person who has lived on this planet for nearly a century will have learned a good deal about how the world works. As the English novelist John Cowper Powys put it, "If by the time we're sixty we haven't learned what a knot of paradox and contradiction life is, and how exquisitely the good and the bad are mingled in every action we take, and what a compromising hostess Our Lady of Truth is, we haven't grown old to much purpose." Powys is describing the wisdom that under the best of circumstances may accompany maturity.

Wisdom has been the lifetime interest of Dr. Paul Baltes of the Max Planck Institute in Berlin. Baltes defines wisdom as "a state of knowledge about the human condition, about how it comes about, which factors shape it, how one deals with difficult problems, and how one organizes one's life in such a manner that when we are old, we judge it to be meaningful."

In a typical Baltes experiment, subjects are asked to think aloud about how a fictitious person might go about solving a specific life problem such as the following:

Michael, a mechanic aged 28 years with two preschool children, just learned that the factory in which he's working will close in three months. There is no possibility for further employment in this area. His wife has recently returned to her well-paid nursing career. Michael is considering the following options: he can plan to move to another city to seek employment, or he can plan to take full responsibility for child care and household tasks. What should Michael do and consider in making these plans?

After hearing this scenario, the elderly subjects in Baltes's study are asked to think aloud about Michael's situation and what he might do about it. Baltes rates their responses according to several wisdom-related criteria. Included here are factual and procedural knowledge about life (what to do under certain circumstances and how to go about doing it); the likely meaning of events; an appreciation of context and the ability to take a long overview rather than a narrow here-and-now approach; acceptance of the inherent ambiguity and uncertainty in any complex situation.

Here is the response of an older man who rates high on the wisdom scale:

First of all, Michael should consider what life is all about. What are his aims? And how is his life related to his job, on one hand, and to his family, on the other? If he has in mind starting a career and he hopes to bring his family later to the place he's moving to, that's one choice. The other choice would be to stick to his family, to take over the role of the houseman, and so give his wife a chance. However, I wonder if his intention to do this would hold if he really had to do houseman's work week after week. Further, in case he did move, would his marriage hold up? I would also ask you for data about his wife, the kids, the social environment in which he is living now. What is his wife thinking? How will the children react if he goes off to another city?

Although not every older person expresses the analytic and emotional sharpness of this respondent, Baltes has found that older persons (average 72 years of age) perform as well on such tests as younger volunteers (average age 32). Further, when it comes to psychological insight, older individuals performed as well as clinical psychologists, who, as a group, are superior exemplars of wisdom. Thus, when it comes to the social problem solving that is an important component of wisdom, there is no reason for us to be discouraged. In Baltes's studies a healthy older person performs at least as well as, and usually better than, his or her younger counterparts.

In addition to wisdom, Baltes also investigated intellectual functioning in the old and very old participating in the Berlin Aging Study (516 people ranging in age from 70 to 103, with an average age of 85). A battery of fourteen tests were used to assess five cognitive abilities: reasoning, memory, fluid intelligence, perceptual speed, and crystallized intelligence (knowledge). Baltes found great variability among the participants—another indicator that people age at different rates and that a general decline is not an inevitable result of aging. Baltes's findings are consistent with a sixteen-year study of twenty-five people between 71 and 95 carried out by John Morris, of the Memory and Aging Project, Washington University School of Medicine, St. Louis, and associate professor of neurology at Washington University in St. Louis. Morris found that some individuals maintain alert inquiring minds well into their 90s and beyond. "If brain function becomes impaired, it's a result of disease, not age," Morris concludes.

Contrary to popular belief, the healthy aging brain does not suffer significant loss of neurons in the cerebral cortex, the very top of the brain. Studies done in the 1950s and 1960s suggested losses of up to 30 percent of cortical neurons in the normal brain. These studies resulted in a dogmatic claim by many writers on the brain (including myself in several of my earlier books) that we lose fifty or sixty thousand neurons per day. More recent and more extensive studies suggest this was an exaggeration. Only modest cortical-cell loss occurs in normal aging.

"For years neurologists were taught, and in turn taught others, that the aging brain suffered a staggering loss of nerve cells," according to Guy McKhann of the Zanvyl Krieger Mind/Brain Institute of Johns Hopkins University. "What's really taking place is a decrease in the number of brain cell connections."

Microscopic examination of nerve cells from the brains of older people shows not so much a diminution in numbers as a thinning out of the *dendrites,* the parts of neurons that act as receivers for impulses from other neurons. Picture in your mind a tree in its full summer bloom, when it has the greatest number of small branches and leaves. Now imagine that same tree in winter, when the branches are denuded and starkly outlined against a slate sky. That transition looks very much like what's seen when neurons from the young are compared to neurons taken from older brains. But does this thinning out and paring down have to happen? Not if studies in rats and other laboratory animals are any indication of what happens in our own brains.

Enrich a laboratory animal's living environment and the brain retains its summer plumage of richly interconnected neuronal connections. This holds true for older animals as well, suggesting the likelihood that it's never too late to alter the dynamic neuronal connections by increasing one's interests and mental activity.

"That old adage 'Use it or lose it' conveyed a lot of wisdom that neuroscience is now proving to be true," says McKhann.

"In addition we shouldn't lose sight of the fact that there are differences in the aging patterns among people. We are not all aging in the same way or at the same rate."

The potential for neuronal cell growth never completely disappears in the normally aging brain. While the dendrites on some neurons are thinning out, the dendrites on other neurons are growing additional extensions, reaching out and touching, thus increasing the opportunity for new synaptic contacts. A balance exists, in other words, between retraction and expansion.

The dynamism of the brain, changing as it does at every moment of our lives, does not cease in the later years. Instead, it retains a plasticity whereby dendrites expand and interconnect in direct relation to our level of intellectual activity and stimulation. Here is how it is put by neuroanatomist Arnold Scheibel: "The concept of two types of neuronal responses to aging, one involving dendritic retraction and one reactive dendritic expansion, brings with it a number of exciting implications for providing more effective and fulfilling lives for the elderly."

Among these implications is the likelihood that the maturing brain remains capable of learning new information and taking up new activities. Remember the old adage "You can't teach an old dog new tricks"? You can, but if you want to teach new tricks, you have to apply new methods, since the mature brain learns in different ways.

"Although we cannot prove that enhanced intellectual activity preserves healthy brain function into old age, it seems to make a good deal of sense. Besides, intellectual activity makes for a richer, more informed life, and that carries its own reward," says Guy McKhann.

• • •

Scientists refer to *robust aging* as a description of the way all of us would like to be as we get older. We want to feel good, enjoy good physical and mental health, remain productive, and stay cognitively intact. Researchers at the School of Public Health at the University of Michigan reported in 1995 their findings of how robust aging differed from just plain aging. They found the most robustly aging individuals had more friends and social contacts, better vision and general health, and fewer significant life events both good and bad in the previous three years. Each of these three factors appeared not to be correlated with the others, which suggests that robust aging is a multidimensional phenomenon.

Certain activities can also be counted on to stave off premature and pathologic brain aging. A three-year study of 2,040 older people living in Gironde, France, highlighted the following activities as associated with enhanced brain function: regular participation in social and leisure activities, such as traveling; keeping busy with odd jobs around the house; knitting; and gardening. None of the subjects in the study were involved in all of these activities (knitting remains a predominantly feminine pastime in France). While participation in any of them increased the likelihood of enhanced brain functioning, the healthiest subjects were involved in several activities.

Obviously, participation in physical and mental activities demands a certain minimum of physical endurance. Muscle building is one sure way of building up endurance; it also combats imbalance and the resultant falls. By the time a person reaches 75, he has lost about half of his muscle mass.

This reduces that person's reserves to the point that any en-forced inactivity (perhaps prompted by the flu or, even worse, a broken bone) may push the person over the edge, to still-smaller muscles and reduced overall muscle mass—factors that increase the likelihood of falls, more fractures, additional inactivity, and even smaller muscle mass.

A 1994 study from Tufts University showed that it's never too late for muscle-building exercises. Even frail elders in nursing homes were able to benefit from a muscle-strengthening program; some improved enough to go from a walker to a cane after only a brief exposure to a low-level training program. A 10 percent addition of general muscle mass as a result of an exercise program leads to improvements of more than 100 percent in strength over a ten-week period, according to the Tufts study.

Muscle-building programs involve the brain as well as the muscular system. The initial event takes place in the brain as you decide, say, to lift a barbell. This intention is conveyed from brain to spinal cord and out along the length of the peripheral nerves serving the arms. Muscles respond to nerve impulses conveyed along the peripheral nerves. At the neuro-muscular junction (the point of contact between nerve and muscle) the electrical nerve impulse is converted to a chemi-cal message. A neurotransmitter, acetylcholine, crosses the gap separating nerve ending from muscle layer and latches onto a specialized receptor. Activation of the receptor stimu-lates the muscle to contract.

With age the number of nerve-muscle junctions (referred to as *motor units*) decreases. With fewer motor units comes less muscle activation, leading to a loss of muscle tissue. For years neurologists assumed that the loss of motor units is an inevitable consequence of aging. Now many suspect that the real culprit may be inactivity rather than aging per se. As a consequence, doctors now believe it is never too late to start a muscle-building program. At any age, muscle building can counter the pernicious cycle of physical inactivity and frailness-related injuries that account for so much disability in the later years. In a study supported by the National Institute on Aging, strength training proved to be the single most important factor in preventing falls.

While bodybuilding contributes to the maintenance of muscle bulk and tone, it may still not be sufficient to combat the loss of balance that afflicts older people who haven't taken preventative measures. For instance, most people experience increasing difficulty balancing themselves on a single leg as they get older. In order to accomplish something like putting on a pair of trousers, the older person often must sit down or seek support by leaning against a bureau or cabinet at the moment when the body weight must be shifted from one leg to the other. There are several reasons for this difficulty.

As already mentioned, strength is decreased in the legs as a result of the loss of muscle bulk. In addition, the brain's balance mechanisms frequently falter as a result of aging changes in the nerve pathways from ear to brain stem to cerebral cortex. Finally, the conduction fibers in the back part of the spinal cord, tethered together in what neurologists refer to as the posterior column, undergo age-related delays and inefficiencies in nerve-impulse transmission. Posterior column fibers convey information about position of toes, feet, and legs. As the posterior columns begin to malfunction with age, the person must frequently look downward to keep track of the movement of his lower limbs.

As a result of all of these influences, the mature brain lags ever so slightly behind in its task of maintaining a dynamic and instantaneous representation of the body's location in space. A mismatch develops between the actual body location and the place where the brain thinks that it is. The difficulty is most pronounced on occasions such as failing to anticipate a curb, suddenly missing a step, and catching oneself at the last split second by a rapid gait readjustment. In that split second the brain must rapidly recalculate the body's position in space, incorporate that knowledge into a dynamic representation, and issue the necessary orders to reposition the legs so as to achieve a position of general body equilibrium. With age and inactivity these calibrations and readjustments take too long, and the correction cannot be made soon enough to prevent a fall.

A promising approach to positively influencing balance and equilibrium is provided by the regular practice of tai chi chuan (also called TCC or Chinese shadowboxing). Tai chi consists of a series of slow, deliberate movements involving trunk and body rotation, combined with periods during which the practitioner stands fully supported on a single leg. Used for centuries in China and other Eastern countries as an exercise for older individuals, this ancient martial art strengthens muscles and improves stability.

In a study published in the *Journal of the American Gerontological Society* in May 1996, Steven Wolf, director of research in the Department of Rehabilitative Medicine at the Emory University School of Medicine in Atlanta, confirmed what elderly Chinese discovered centuries ago. Tai chi practitioners are much less prone to falls and resultant fractures when compared to people who do no balancing exercises.

In one study of two hundred people 70 years of age or older, those participating in a fifteen-week tai chi program reduced their rate of falls by 47.5 percent. This is a critical reduction, since about one-third of older people suffer a serious fall each year, with close to 25 percent sustaining a disabling or life-threatening injury, such as a hip fracture or subdural hematoma (blood clot) on the brain. Not only can tai chi help to prevent these injuries; this ancient exercise can also reduce the incidence of something equally disabling: the fear of falling.

"Fear of falling is probably one of the biggest concerns among the elderly population," claims Adam Karp, medical director of the Geriatric Falls Prevention Program at the Hospital for Joint Diseases Orthopedic Institute in New York City. As a result of their fear of falling, many older people refuse to leave their homes and apartments, further isolating themselves from the social contact they require. In the study carried out by Steven Wolf in Atlanta, participants who had practiced tai chi for fifteen weeks were far less fearful of falling than they had been at the start of the study.

Tai chi also confers cardiorespiratory benefits, and when practiced regularly "may delay the decline of cardiorespiratory function in older individuals," concludes a study, "Two-year trends in cardiorespiratory function among older Tai Chi Chuan practitioners and sedentary subjects," published in 1995 in the *Journal of the American Geriatric Society* by a team of rehabilitation specialists at the National Taiwan University Hospital, Taipei, Republic of China.

While tai chi is an exercise, it is also a form of meditation (the Chinese refer to it as a "mindful exercise"). But instead of sitting, as in most forms of meditation, the tai chi practitioner meditates via total absorption in slow, rhythmic body movements and gentle breathing patterns. According to results from the Exercise Physiology Laboratory at the University of Massachusetts Medical School in Worcester, mindful exercise with tai chi can reduce tension, depression, anger, and total mood disturbance—all brain-based disorders.

As with many ancient and venerated rituals, tai chi can involve the memorization and carrying out of complex and off-putting "forms" on the part of the aspiring practitioner. Several of the forms commonly taught consist of over a hundred different foot and hand positions. Obviously, not everyone intent on improving balance and coordination may be interested in involvement at this level of detail.

A more sensible approach to the exercise was summed up by one of my patients, an 85-year-old Chinese man: "Don't worry about learning any of the forms, at least not any of the longer ones. Instead, just concentrate on repetitively shifting your total body from one foot to the other. While doing this, let your hands move easily and gracefully in patterns that come naturally. Do not think too much about what you're doing. Just build up your ability to bear the body weight on each leg without swaying."

• • •

But of all the factors contributing to a successful brain functioning in later years, education ranks highest. A fifty-year follow-up of elderly male twins first tested as armed-forces inductees in the early 1940s showed that cognitive performance in early adult life predicted cognitive abilities in the later years. Those with the highest measures of intelligence and most extended education showed the least cognitive deterioration with aging. It's likely that more education, higher intelligence, larger brains, or more synaptic contacts provide a greater cognitive reserve, thus allowing the mature brain a greater tolerance for injury.

"Education leads to more synaptic connections; it takes a longer time to lose these greater number of connections," according to Zaven Khachaturian, director of the Alzheimer's Association's Ronald and Nancy Reagan Research Institute. To make his point during our discussion, Khachaturian directed my attention to several trees beginning to lose their leaves outside the restaurant window where we sat. "A few months ago all of those trees had more leaves than they do now. But the trees didn't all have the *same* number of leaves. Some had more leaves to start with and therefore had a greater reserve capacity. If you think of the leaves on the tree as a metaphor for neurons in the brain, then trees with an increased number of leaves to start with will age slower and better as winter approaches. This, of course, assumes that all of the trees are losing leaves at the same rate, which is probably not true any more than it's true that all brains lose neurons at the same rate."

Extending Khachaturian's analogy a bit further, the branches of the tree and the richness of their patterning correspond to the number and complexity of dendrites and their dendritic spines, the knoblike extensions on the dendrite that play an important role in receiving messages from other neurons.

Education and intellectual pursuits increase the number of synaptic connections, with the greatest number of synapses resulting from the most intense and prolonged learning. "Education creates more synaptic connections and enriches the dendrites. Since it takes longer to reverse these changes, those with a better education stay mentally alert longer than those with less learning," according to Khachaturian.

A good deal of experimental support exists for Khachaturian's hypothesis. Rats provided with toys and other enrichments end up with bigger brains as a result of increases in the size of neurons and their numbers of synaptic connections. Most of the experts on brain aging agree with the proposition that education leads to an enriched brain. But there is a conundrum here: Does a person keep his brain young by prolonging his education and remaining intellectually curious? —or do education and intellectual curiosity serve as markers for a brain that is already enhanced and, for unknown reasons, continues into advanced age to show more adaptability, neuronal plasticity, and dendritic arborization?

While no one has a definitive answer to this chicken-and-egg question, it's likely that the brain, at whatever level of endowment it starts, builds up a reserve capacity during the early learning years. This provides protective action against the losses of normal and abnormal aging. For example, neuroscientists now know that, in the later years, neurotransmitters and hormones, particularly from the hypothalamus, gradually lessen in concentration or produce fewer effects. Depending on the brain's reserve capacity, this falloff may occur earlier or later and may vary in degree from one person to another.

Continuing education and learning enhance the brain's reserve capacity through their effect on brain structure and function. Whenever you have learned a new skill, you have done so by modifying your brain's organization. Each time that skill is practiced, the brain circuit is facilitated—the

reason athletes and musicians practice and train so assidu-
ously. But once you stop practicing that skill, even for brief
time periods, the neuronal circuit becomes less facilitated; the
skill is harder to perform, and the performance is less skillful
and is accomplished less successfully. Practicing our mental
skills compensates for aging changes by not allowing disuse
to occur. Practice also leads to the buildup of reserve capacities either by maintaining long-established neuronal circuits
or by establishing new ones. Finally, practice can lead to the
development of novel strategies for maintaining performance
levels.

Thus, the greater the education during the formative years,
the greater protection later against the inroads of brain disease and aging. While this may not seem entirely fair, since
not everyone has the opportunity (or the ability) to prolong
his or her education, it should hardly come as a surprise.

Considerable variation exists between people when it comes
to their general health and well-being. For instance, people
differ markedly in muscle strength, lung capacity, and general physical endurance. Why would the situation be any
different with the brain? No one doubts anymore that, irrespective of educational opportunities, some people are more
intelligent than others. Could those differences in intelligence
be the result of a more enhanced and adaptable brain? If so,
then it seems reasonable to assume that such a brain would
seek out opportunities to learn more and continue the educational process as long as possible. If this is true, then a self-
selection process is at work here: more evolved brains are
more curious about the world and thrive on knowledge. But
even this rather deterministic view does not shut the door on
the positive effects of education in people with brains not so

well endowed. The operative words here are "curiosity" and "stimulation."

Enhanced dendritic connections can be created by education. To employ Khachaturian's analogy, educational stimulation causes more leaves to remain on the tree for a longer time. Two brains with different numbers of leaves (that is, neurons and neuronal connections) will age very differently according to the amount and quality of stimulation they encounter. Of course, we all have influence over the stimulation in our lives. We can seek experiences that promote mental and physical stimulation. In the process we are contributing to the enhancement of our brain.

• • •

It's wonderful news that education is a particularly important protector against mental decline. That's because education, like physical exercise and social activity, is an area of our lives over which we retain a good deal of personal control. While we have no control over our genetic inheritance, and imperfect control over our health, we can further our education and learning.

The Shimamura study referred to in Chapter Two concentrated on formal education, that is, the number of years spent in school and the highest level of education achieved (as evidenced by diplomas and academic degrees). But education doesn't have to involve degrees and time spent in classrooms. Cultural experiences at any age can compensate for limited educational opportunity earlier in life.

For example, adult education courses are widely available. Such activities will establish new brain circuits and enhance existing ones via concentrated, focused attention on an area of study, memorization of an assigned amount of material, organizing and assimilating what one has heard or read, integrating the learned material into one's personal mental repertoire, and responding to tests and questions about what has been learned. With each of these processes the brain is reshaped and rejuvenated.

Thus, education is a lifetime process that through brain modification confers antiaging benefits. What's more, experience enhances the sense of well-being: as our brain's performance improves as a result of our learning, we generally feel more empowered and more confident when, as a result of our educational efforts, we achieve mastery over some body of knowledge.

Nor does education necessarily have to involve traditional book learning. It can range from learning how to coordinate the preparation of a gourmet meal to the management of a stock portfolio. Discussions, widespread interests, curiosity, and the desire to learn as much as one can about a wide-ranging series of topics can substitute for years of formal education. Many entrepreneurial successes fit into this category.

Typical is an acquaintance of mine with an eighth-grade education who has made himself a millionaire as a result of studying real estate and the stock market. Since these two areas of interest involve economics, psychology, international and national trends, financing and accounting, history and sociology, he has read widely on each of these subjects. Although his knowledge is uneven and idiosyncratic (as is typical of the self-educated), he can usually hold his own with people his own age and younger who have a decade or more additional formal education. He serves as living proof that the benefits of education aren't dependent on the schools, the classmates, and the teacher (although all three can contribute) but on the level of brain stimulation achieved.

• • •

Marilyn Albert of Massachusetts General Hospital and Harvard Medical School has tested thousands of older people over the previous two decades. She has discovered *four factors associated with enhanced brain function and healthy brain aging:*

- *Education* heads the list. Education produces greater and more elaborate neuronal connections, as discussed above.

- *Strenuous exercise.* Increased physical activity increases the blood and oxygen supply to the brain by maintaining the health of the blood vessels in the path from heart to brain.

- *Enhanced lung function.* Healthy lungs, as measured by greater pulmonary capacity with exercise, increase the amount of oxygen carried in the bloodstream from the lungs to brain.

- *Self-efficacy.* This is defined as a belief in the worth of one's accomplishments and a belief that one is in control of one's future health and destiny. Self-efficacy is a measure of how stressful circumstances are understood and responded to. At all times stress varies according to the interpretations a person applies to a situation. One person's stress is another person's challenge. If a person feels helpless in the face of uncontrollable stress, that person's adrenal glands will put out increased levels of glucocorticoids, the stress hormones. These increased levels directly damage the hippocampus on each side of the brain. We know for certain that this happens in animals and can reasonably infer that it happens in human brains as well.

Of Albert's four factors, self-efficacy, basically the feeling of being in control of one's own life and destiny, may be the most important. People differ considerably when it comes to the feeling of being in control. Moreover, these beliefs have consequences.

A sample of 682 people aged 70 years and over living in New Zealand revealed that those with a feeling of a great deal of control, the "believers," feel generally good about themselves, are less likely to be depressed, drink less alcohol, and take in less protein. In terms of physical health, the believers rarely suffer from chronic obstructive lung disease, are more satisfied with their bowel functioning, and are more likely to participate in moderate to strenuous activity. You will notice that disturbances in several factors on this list are frequently encountered among depressed people. Depressed people drink more, and the increased alcohol worsens their depression. Many depressives also experience chronic constipation and bowel irregularity. In fact, the worse the depression, the more likely it is that the depressed person will exhibit one or more of these disturbances. Based on this observation, psychiatrists think it likely that depression and depression-related factors play a large part in determining a person's sense of control.

Although good physical health figures prominently in both

Albert's and the New Zealand findings, other aspects of life can make up for poor health.

In a study involving individuals in Hong Kong 70 years of age and older, the most important factors contributing to a healthy brain and "life satisfaction" were support provided by family members and adequate income to meet living expenses. Although the culture in Hong Kong is much different from that in the United States, some of the other components of the life-satisfaction index are quite intriguing. Factors associated with the highest degree of life satisfaction included higher social class, educational attainment, self-perceived good health, good hearing and vision, daily exercise, absence of recurrent falls, a low incidence of depression, satisfactory living arrangements, good social support, and participation in social activities. Most important is the maintenance of social interactions.

According to researchers at the Max Planck Institute for Human Development and Education in Berlin, successfully aging people anticipate the loss of friends as they grow older and make deliberate efforts to make new ones, thus enhancing what the researchers call feelings of social embeddedness.

Stamina (increased staying power in the face of misfortune; ability to change and function effectively when stressed) also plays an important role in robust aging. For the most part stamina in the later years is the result of successful adaptations earlier in life.

"Stamina in later life is contingent on a triumphant, positive outlook during periods of adversity," according to psychologist E. J. Colerick. "Persons who view situations involving loss as threatening, overwhelming and potentially defeating, experience no such outcome; low levels of stam-

ina mark their later years." Colerick emphasizes the importance of attitude and positive orientations as a determiner of
stamina.

Closely related to stamina is resilience. In a nursing study
of successfully adjusted older women, five underlying traits
were identified as contributing to resiliency. The successfully
adjusted women were more self-reliant, more tolerant of
being alone, more perseverant in the face of stress. In addition
they maintained an abiding sense of equanimity and, in general, considered their lives as meaningful.

Courage is another important trait that reappears again
and again in studies of robust aging. According to D. L. Finfgeld, who interviewed people ranging in age from 67 to 94,
"Becoming and being courageous involves problem solving,
the transformation of struggles into challenges, a sense of
personal equanimity and personal integrity."

Nineteenth-century American history provides examples of
the longevity benefits of keeping oneself challenged. While
the average woman in the Civil War era lived to the age of
about 40, a group of extraordinary nurses—including Louisa
May Alcott, Dorothea Dix, and Clara Barton—survived to
much older ages (Barton to 91). While many explanations
have been offered to account for their increased longevity
(altruism, religious faith, social and marital status), the most
important factor, according to nursing historian W. Woodward, was the presence of a "pioneering spirit."

Nor is that pioneering spirit a thing of the past. While
working on the *Mind* television series in 1988, I personally
encountered a contemporary example of that spirit in Hulda
Crooks, who at 91 became the oldest woman to climb Mount
Fuji. It's likely that the pioneering spirit acts as a powerful

motivator for the older person to accomplish more and thereby stay alive as long as possible.

Cicero would not have been surprised with our new knowledge about the factors promoting a healthy old age. He anticipated several of these factors more than two thousand years ago:

But it is our duty . . . to resist old age; to compensate for its defects by watchful care; to fight against it as we would fight against disease; to practice moderate exercise; and to take just enough of food and drink to restore our strength and not to overburden it. Nor, indeed, are we to give our attention solely to the body; much greater care is due to the mind and soul; for they, too, like lamps, grow dim with time, unless we keep them supplied with oil . . . intellectual activity gives buoyancy to the mind.

5

ADVICE FOR SUCCESSFUL AGING

*D*uring research for *Older and Wiser*, I interviewed many people who are aging successfully and asked them what advice they would give to others. My interviewees were all 70 years of age or older and still productive. I asked that their advice not involve obvious impracticalities for the average person.

My subjects included

Art Buchwald (73), nationally syndicated columnist and author of many books, including his autobiographical works *Leaving Home* and *I'll Always Have Paris.*

Morris West (80), author of twenty-seven novels, including *The Shoes of the Fisherman* and, most recently, *Vanishing Point.* A past recipient of the Royal Society of Literature Heinemann Award, he is currently at work on the fourth novel in his *Papacy* series.

Harriet Doerr (86), winner of the National Book Award for her novel *Stones for Ibarra.* She started writing at age 65 and, despite a severe late-life-onset visual handicap, continues to produce novellas and short stories. *The Tiger in the Grass: Stories and Other Inventions* was published in 1995.

Charles Guggenheim (71), documentary filmmaker and winner of four Oscars during his career. His works include *Nine from Little Rock, Robert Kennedy Remembered,* and *A Time for Justice.*

C. Vann Woodward (88), one of the nation's most respected and prolific historians. Now retired from a teaching career at Johns Hopkins and Yale University, he serves as the general editor for the *Oxford History of the United States,* published by Oxford University Press.

Irving Kristol (76), cofounder and editor with Stephen Spender of *Encounter* magazine, a fellow of the American Academy of Arts and Sciences, and currently coeditor of *Public Interest* magazine and publisher of *National Interest* magazine. He is also a director of the Dreyfus Money Market Instruments, Inc.

Daniel Schorr (80), journalist and commentator for National Public Radio.

Olga Hirshhorn (75), one of the world's most famous art collectors. Her *Olga Hirshhorn Collects: Views from the Mouse House* was featured in 1995 at the Corcoran Gallery of Art in Washington and served as a preview of the seven hundred works in her collection that she has bequeathed to the Corcoran.

Chalmers M. Roberts (85), chief diplomatic correspondent for the *Washington Post* until 1971. He continues to contribute occasional pieces to the *Post* and published in 1991 *How*

Did I Get Here So Fast? Rhetorical Questions and Available Answers from a Long and Happy Life.

From these interviews I extracted ten factors mentioned most often by my subjects as keys to successful aging. I then related these factors to what we know about the brain. To my great satisfaction I discovered that the factors mentioned by my subjects as promoting good mental and emotional health correlate with the factors now known to be important in the preservation of optimal brain functioning.

Here are the ten factors for healthy brain functioning: (1) education, (2) curiosity, (3) energy, (4) keeping busy, (5) regular exercise and physical activity, (6) acceptance of unavoidable limitations, (7) the need for diversity and novelty, (8) psychological continuity over the life span, (9) the maintenance of friends and social networks, and (10) the establishment and fostering of links with younger people. Here, in their own words, are the recommendations of these stellar exemplars of robust aging. In several instances I have supplemented their comments with reflections on how their recommendations enhance optimal brain functioning.

Education

CHALMERS ROBERTS

When you boil it down to the essentials, that means keeping your heart pumping, your noodle active, and your mood cheery. One theory is that the best way to stimulate your mind is to receive and absorb a constant supply of information.

Keep the gears moving, so to speak. Don't just read the morning newspaper. Get out of the house. Lunch with friends. Go see these movies the young describe as "awesome." Read something that challenges the conventional wisdom in whatever field holds your lifelong interest. Travel is tremendously stimulating, but you have to do your homework. I always try to do my essential reading before leaving home. This wisdom is carved into one of the massive panels at the top outside front of Washington's Union Station: "He that would bring home the wealth of the Indies must carry the wealth of the Indies with him. So it is in traveling—a man must carry knowledge with him if he would bring home knowledge."

CHARLES GUGGENHEIM

The more education you have in terms of knowledge and curiosity about the world, the more you are able to further your enrichment. Education enables you to extend your life into periods that preceded you. A familiarity with history, for instance, gives you an opportunity to live in any century you want and live the life in your imagination of anybody who has ever lived before. That's what education and history are all about. And that's why it's so sad to see older people retiring to a life of golf or mindless travel or incessant "busyness." There is an opportunity out there, a late-life chance to deepen and learn. That's why we can never stop learning or stop asking questions. I often tell people who neglect the opportunity for additional learning and education that they are living in a house with five rooms that they have yet to visit.

• At any given moment, any one of the 10 to 20 billion neurons in our brain is potentially capable of establishing connections with any other. Thanks to elaborate networks, no cell is more than three or four degrees of separation from another. What's more, the brain thrives on the number and richness of its connections. The take-home lesson from these facts seems obvious: learn more, learn something new, and the brain's network of interconnections is further elaborated.

It's incredible to me that although everyone would agree about the inadvisability of consigning a high-performance car to inactivity for long periods of time, few people realize the same principle holds for the highest-performance vehicle in the universe.

Education is an enriching and liberating process. Knowledge brings joy—a joy you can see on the face of a fourth-grader who has just successfully solved a math problem involving fractions, or on the face of an 80-year-old learning about the Crusades in an adult education course. The brain is designed to process knowledge and information just as the digestive system is designed to process food or the lungs process oxygen. If food, oxygen, or knowledge is cut off, the organism dies. It's that simple.

Curiosity

This is the mental trait mentioned most frequently.

ART BUCHWALD

To remain creative and mentally sharp, you have to come up with things nobody else has thought of, or you have to deal with familiar things in novel ways. But most important of all, you have to have a sense of curiosity. For instance, just a few minutes ago while walking here on the Vineyard [Martha's Vineyard, where our discussion took place], I came upon a minor traffic accident. I stopped because I was curious about the people involved, and how the accident happened and why it happened. Other people walk by these kinds of situations and think only about such things as insurance, who was at fault, and who might have to pay. But that is no way to remain sharp and mentally vibrant. You have to stay curious about people and what they are doing. Who are the people in the cars? Where were they going? What is their response to this relatively minor mishap? I'm curious about such things. If interest and curiosity stop coming automatically to you, then you're in trouble, no matter how young you are. In fact, I believe interest and creativity prolong life and enhance life. I believe people who are interested in the people and events around them live better and feel younger.

• The importance of curiosity to retained optimal mental functioning should come as no surprise. Curiosity is an integrative and complex function related to motivation, arousal, attention, and preference for novelty. All of these depend upon the operation of the brain at its highest levels of performance. Moreover, curiosity can and must be cultivated at every age. Not only is it a marker for healthy brain aging, but it's likely that the deliberate cultivation of curiosity protects against brain degeneration. As we become more curious about the world, our interaction and involvement with others increases as well. This leads to even more information, and hence increased curiosity. The more we learn, the more we want to learn.

Energy

CHARLES GUGGENHEIM

One of the challenges in the later years is to generate the same amount of energy you did when you were younger. I discovered the secret of how to do that while working on my film about Harry Truman. Truman, as well as Lyndon Johnson, a subject of another documentary film of mine, took regular naps. Truman napped for a half hour after lunch. Johnson went even further and put on pajamas before retiring for his afternoon nap. When Johnson woke up he started another eight-hour work session. The nap enabled him to put in two eight-hour workdays in the same twenty-four-hour period. On my 71st birthday I decided that if Truman and Johnson took naps, maybe I should take a nap. This decision took some self-convincing, incidentally. Our culture frowns on naps as unproductive. How many people do you know that will tell you without a mild sense of embarrassment that they take naps? Nevertheless, I decided to try napping and see what happened. Now, after lunch, if I feel in a low-energy state, I take a short nap. Like Truman and Johnson, I wake up after twenty or at most thirty minutes feeling energized and refreshed. But interestingly, I don't have to take the nap. If I am deeply involved in something or if I have something on my mind I want to do, I have no need or desire to nap and I skip it for the day. What's more, I don't miss it at such times. A nap is a marvelous means for neutralizing that little falloff in energy that comes in the later years.

———

• The key point to remember: with the passage of the years, some diminishment in energy reserves is normal, even inevitable, due to loss of activating and alerting neurotransmitters from the subcortical nuclei (the "juice machines"). The first step toward overcoming this loss of energy is to realize that the problem does not involve a loss of skills or abilities but simply a diminishment in raw energy. While this loss of energy can be combated by stimulants, sometimes the best approach to replenishing one's energy reserves is a nap or quiet rest.

As Charles Guggenheim remarks, the nap is restorative and can easily be dispensed with at times when external events provide the impetus for renewed energy. This ties in with the curiosity principle: the more involved and enthusiastic you are about people and events, the more energized you will be.

Keeping Busy

OLGA HIRSHHORN

After my long second marriage and when I became a widow at 61, I knew I had to make a life of my own. Nobody was going to make it for me. I had to become a person unto myself, and to do that I had to work at it.

First, I determined to keep physically busy. Every morning I ride my bike. I play tennis regularly and some occasional golf. And it's never too late to take up something new. I began skiing at 60 and horseback riding at 65. I was nearly 70 when I took adult sailing lessons at the yacht club here in Vineyard Haven.

For mental exercise I read a lot. I keep two books going, one in a bag I carry around with me during the day, and the other by my bed. I also keep a date book—I can tell you where I had dinner on this date thirty years ago. Three years ago I started bridge lessons to keep my mind keen. One or two evenings a week I get together with friends to play Pictionary, backgammon, dominoes, or bridge.

I also think it's important to remain open to new experiences and spontaneous proposals. For instance, a year or so ago a friend I knew from the art world suggested at a dinner party that I go with him on a trip to India. I went and experienced a part of the world I had only read about until then.

Overall, I've found that the more I do, the more I want to do. My activity keeps my brain working marvelously well. As an added bonus, I don't have time to brood about the past or get depressed.

DANIEL SCHORR

External stimulation is very important to me. And fascination with news events going on in the world provides me with my greatest stimulation. I guess you could say I am a news junky. I wake up in the morning and put an earplug in my ear (so as not to awaken my wife) and then listen to the news. After that, I get up and read four papers thoroughly. Two days a week I call in NPR and tell them what news story I will comment on that day. Writing the commentary provides the framework for my activities at least up until lunch. Following this, there are then any number of business calls and things to be done at home or at the office. Obviously, I don't like a lot of empty time. Over the years I've found out empty time isn't good for me. In 1965, for instance, I suffered from an ulcer. Almost invariably I would get an ulcer attack when I went on vacation. This pattern is repeating itself now with an episode of shingles I've recently been bothered with. I experience no discomfort while writing a commentary, but as soon as I'm finished I begin to feel the pain again. This is like my ulcer all over again, I'm thinking. The more things I do that I enjoy doing, the better I feel. In fact, I'm only tired when I don't have anything to do. Instead of tiring me out, one activity generates excitement and energy for another.

Regular Exercise

CHALMERS ROBERTS

By our 80s most of us have to make some concessions to the state of our physical health. We do need exercise, and my regular exercise is swimming, skinny-dipping when we have the pool to ourselves. There's nothing better. One writer I know rhapsodizes, "One of the sport's main attractions is its Zen of quiet, meditative tranquillity—back and forth, back and forth—that lets the mind float off to peaceful levels of creativity and well-being."

Acceptance

HARRIET DOERR

In my old age I am afflicted with dreadful eyes. The doctor says I'm functionally blind now. Indeed, I am thankful I can see at all. My right eye has glaucoma and retinal problems.

In my left eye I have no central vision but only peripheral vision. When the glaucoma doctor's examination confirmed the loss of my central vision, he said, "Don't belittle peripheral vision. That's how we see the tiger in the grass." Then he added, "It's also how the tiger sees us." I remembered that remark and used it as the title for my latest book.

Adjusting to my loss of vision has been hard, but I have enjoyed some secondary benefits. When you can't see, you think a lot more. The best exercise of my imagination is sometimes lying in bed a few moments in the morning in the silent house and letting things like a difficult sentence or the ending of a story simply float into the mind.

I've also learned to employ various methods of substitution for my visual handicap. Over the last seven years I have switched to a word processor. Mine has enlarged black-and-white keys, and I wear special glasses when I'm writing with it so that I can see from farther away.

While I can still see well enough to read with my special glasses, it takes a lot of concentration. But that's not a problem if I'm interested in something. When interested, I can concentrate no matter how long it takes me. Determination is what makes that possible. Determination has a lot to do with creativity. I just make myself do something I want to do, and if it's something interesting, I don't feel tired.

MORRIS WEST

My health isn't perfect. I've had cardiac surgery and must take pills to keep me from cardiac failure. But the pills work well and I can write with reasonable normality and continue the extensive travel schedule that I have always enjoyed. To accommodate to the changes brought on by the years, I've adopted a rhythm for my life. I work a half day till one o'clock, when I have a light lunch, and then rest and swim in the afternoon. At five-thirty or so I have a drink. So far this schedule has worked pretty well for me. I am 80 years old, with most of my faculties still intact enough for me to continue a productive writing career. After finishing my twenty-seventh and latest novel, *Vanishing Point,* I started writing the fourth novel of my *Papacy* series. I have every hope that I will be spared sufficient time to complete it.

ART BUCHWALD

Sleeplessness is something I made peace with years ago. Because I have a form of sleep apnea, I wake up often about four in the morning. Instead of brooding or fretting about my insomnia, I get up and read the papers. Eventually I get tired again and go back to bed and asleep.

––––

• Acceptance demands a balanced approach to life that not every older person achieves. On the one hand, there are the inevitable changes in physical and mental functioning that accompany getting older. One can rage and fret against such things as decreased eyesight or difficulty falling and remaining asleep. Or these undesired changes can be accepted and incorporated into new behavioral patterns. When awakened at night, Art Buchwald reads background material that can be used later as the basis for one of his columns. Morris West works with his heart condition by adopting a regular, physically healthy daily regimen that enables him to stay refreshed and creative. Such approaches not only make good practical sense but are in the best tradition of keeping the brain free from the twin demons of depression and suicide.

Indeed, I'm convinced that failure to achieve a healthy acceptance of the life changes accompanying aging lies at the bot-
tom of many later-life depressions and suicides. Simple denial
doesn't work. One must recognize these later-life limitations
in order to overcome them. Yet at the same time, the older
person must be careful not to sink into fatalism and despair
by too ready an acceptance of things. Perhaps the best sum-
mation of desirable acceptance as opposed to simply giving
up and becoming too accepting is contained in the Serenity
Prayer: "Lord, give me the strength to change the things I
can change, the humility to accept the things that cannot be
changed, and the wisdom to know the difference."

Diversity and Novelty

Morris West

To me variety is the most important factor in retaining good
brain function in the later years. You have to work at maintaining as diverse a range of activities as you're capable of. In
my own case, I have taken up painting and jokingly tell people I hope to become a second Rembrandt. This interest in
painting dates back to some modest art instruction I had in
my youth. Now I have returned to that early interest for
cognate reasons: it keeps me mentally challenged and it enables me to experience the world differently.

I tell others of my age that if they wish to remain mentally
sharp, they must develop real interests, not just hobbies. I'm
talking here about things people can do and achieve. For
some, that may mean joining an association that enables

them to perform a living function, whether that be teaching retarded children or growing roses. The scale of the activity doesn't matter. What's important is that the interest involve stimulation, things to do and achieve, a new focus of interest. The more such activities that people can address themselves to, the better.

IRVING KRISTOL

As you enter middle age, it's time to get started on what I call a serious hobby: something that you enjoy and that keeps you stimulated but yet is not something that you expect to make money from. In my own case I started playing around with the stock market. Over the past twenty years or so, the activity has proved interesting and informative. I've picked up some astonishing information from reading analysts' reports and discussing business and economic trends with asset managers. Nor is it necessary to wager a lot of money. Sometimes I just make the purchases and sales in my head and check in a month or so to see how I would have done. But, of course, that's not half as much fun as actually putting some money down to back up your hunch. If you like gambling but prefer something in which the odds are not weighted against you, then the market provides a nice combination of intellectual challenge and stimulation, some excitement, and even a chance to make a little money.

Linkage and Continuity

CHARLES GUGGENHEIM

I don't think I will ever retire in the conventional sense, but if I did, I would revisit many of the places from my childhood. I made a start on that project three years ago when I took the same train and stopped at the same station in the small town in North Dakota where as a third-grader I had lived happily for a summer with the family of a teacher. The teacher, Ada Louise Carpenter, who lived in our house in Cincinnati, had persuaded my parents to let her take me home with her to North Dakota that summer in order to help improve my reading (I was dyslexic). During my visit three years ago, I just walked through it all again: the prairie dogs, grasshoppers, the long, hot, dry days. I could have spent days just looking at the profile of the landscape and having my senses renewed. It was an opportunity to relive. It was like reading a good book for the second time.

The past can do this for us if we retain an interest in it. This is why historic preservation is so important. Historical buildings represent an age and period of time and thus are capable of transporting us to another experience, which is enriching and informative and exciting. An historic building or landmark reminds us, even when nothing is happening at the moment, that something important once did happen. By retaining links with the past, we live again through the exercise of our imagination. That's why the destruction of a historic site is so tragic. The personal and collective linkages are destroyed and we are all diminished. Somebody has taken something away from us we can never recover again. It's so important to retain our linkages with our personal and collective past. I'm convinced of the value of the past in retaining our mental vitality in the present.

HARRIET DOERR

It's wonderful to return later in life to things you loved as a child. For me that love involved reading and writing. I read all the time as a child. I remember in sixth or seventh grade I would come home to the living room after school, bring ginger ale and graham crackers, fold myself into a chair, and, undisturbed, read *Les Misérables* for hours among my crumbs. In terms of writing, I always liked composition days at school, when I could write about anything that interested me.

After going off to college, first at Smith and then later after transferring to Stanford, I married early and as a result had to wait forty-five years to obtain my degree from Stanford. I returned there at age 65, three years after my husband died. I went back because I was curious and interested in learning new things. I also did it because it seemed like fun.

At Stanford, I signed up for the writing course without any thought at the time that published books would ever come of it. I loved the writing class and the discovery, thanks to the encouragement of my writing professors, that maybe I could put together some beautiful sentences. I experienced an enormous feeling of happiness while writing.

Sometimes people ask me if I am sorry I didn't write earlier in life, but the time then just wasn't right. I know I could never have written amid the hurly-burly of married life, with children and all the other responsibilities. I need quiet time for things like simply sitting in a lawn chair and thinking. The subconscious is very important, and you can't force things. The subconscious is always there, like a faithful companion, but you have to treat it well. I don't try to force myself. I avoid set routines and never write by a clock or make myself sit for an hour waiting for a word or image to come. And I always stop at a place where I still have something to say.

At my age I still care very much about words, and if I come across something wonderfully expressed, that experience is a total show stopper. Good writing will always do that for me. It's wonderful to find good writing on a page written by anyone. If that writing is my own, I feel an enormous satisfaction.

It is never too late in life for anyone to begin writing who shares that enthusiasm for excellence and beauty and yearns just to write a sentence that he or she can be proud of.

Friends and Social Networks

ART BUCHWALD

I care very much about what other people think about me. I feel most myself when I have done something successful and other people agree that I have.

CHARLES GUGGENHEIM

Assume that for some reason you are confined to the town or city you are now in and are not permitted to go outside it. You are not allowed to experience anything outside of this place, which will be your world for the rest of your life. Let's also assume in this mental exercise that all knowledge has been taken away from you of your relatives—your parents, grandparents, cousins. Obviously, not even a jail is that confining. Part of the sense of confinement in our imaginary situation would come from the fact that your life consists of more than yourself and your immediate surroundings. You have been enriched by your connections with lives other than your own and places different from your present location.

Maintaining Links with the Young

C. Vann Woodward

I am unhappy unless I am writing. One important component of that process is staying in contact with my students. In fact, several are coming with their families to visit me over the upcoming weekend. Our interactions are of mutual benefit. They seek my thoughts and suggestions about their work. But I gain too, since being in touch with them enables me to remain creatively in contact with the thoughts of younger people. I really believe I gain as much from my former students as they do from me. Thus I never feel competitive with them or exploited by them. Indeed, I think of my former students as family and friends.

At home I have what I call a "vanity shelf" of books published by my former students. Most of these books started out as published dissertations that I directed. Several of the books achieved great success, such as the Pulitzer Prize–winning

Battle Cry of Freedom by James McPherson. I am happy for such successes and feel happy that I have had the opportunity to be a part of them. That, of course, would not have been possible had I not made deliberate efforts to remain in touch with younger people.

• • •

On the subject of retirement, several of the interviewees underscored the need for never really retiring in the conventional sense.

CHARLES GUGGENHEIM

The important point is that if you choose to retire, you should retire *to* something, not *from* something. I did a film for AARP, and what impressed me the most about the people we interviewed prior to retirement was that few of them had planned ahead more than a few weeks. If asked about their plans, they would mention intentions to clean or refinish a basement—projects that would be over in weeks, at the outer limit—and then what would they do with their time? They hadn't a clue.

ART BUCHWALD

The mentally creative person never thinks of himself as retired. That's because if you keep doing something interesting as you get older, you're a more interesting person, not only to everybody else but yourself as well. One other benefit: if you can continue to keep doing something challenging as you get older, your natural tendency will be to think and feel you're not getting older.

MORRIS WEST

The problem with retirement stems from the absence in our society of the traditional family and tribal relations. The support systems that were the norm in previous societies are no longer present. Neither the parish nor the synagogue function any longer as effective means of bringing people together. This means the aging person is forced to accept so much more responsibility for his own well-being than was previously the case. This isn't necessarily all bad, of course, but it should make a person think carefully about removing himself from the social network he enjoys at work. Finding another "place" for himself, psychologically speaking, may turn out to be more difficult than he imagined.

DANIEL SCHORR

If there is any message I would give to someone considering retirement, it is this: retire to some other occupation or don't retire. Arrange your life in such a way that people call on you or need you for something. That way you remain vital and a part of things. In my own case, I feel tremendously lucky because I enjoy my work and have no immediate intention to retire. And I take all of the necessary steps to make sure there will always be things for me to do that I will enjoy and that will make me feel good.

THREE

6

ALZHEIMER'S

AND OTHER DEMENTIAS

*D*ementia" is a technical term for what's commonly referred to as senility. The earliest forms of this disorder are often so subtle that the condition isn't recognized. A gradual loss of interest in life, mild disorientation in place and time, carelessness about personal appearance and hygiene, personality changes—any or all of these signs of deterioration are inevitably present at some point as the disease slowly evolves. Alertness and conscious awareness, however, are preserved until very late in the illness.

Neurologists distinguish between dementias that seem to occur suddenly, after something like a head injury or a heart attack, and the more chronic forms, which take months or even years to develop. Typical of this later form of the illness is Paula.

• • •

An 84-year-old woman living alone, Paula was brought to me by her daughter because of irresolvable conflicts and arguments they had had over the previous six months concerning the older woman's handling of her finances. Her checkbook no longer balanced; outstanding bills lay scattered on the dining room table; and her bank statements reflected varying amounts of money paid to a general contractor who had so far failed to deliver on house repairs for which he had been fully paid in advance.

When asked about her daughter's allegations, Paula burst into a tirade of abuse. She told me her daughter was interfering in her life, wanted her put away "in a crazy house," was making up "stories" about her. She said the contractor was "a nice man" who was now "accepting" sums of money from her for "investment."

Rather than take a side in the conflict, I confined my initial remarks to relaxed casual comments about the weather and certain topics currently in the news. Paula's responses were pleasant enough but vague. When I asked her more specific questions, such as the date and our present location, she said: "Things like that don't matter anymore. I no longer work, so why should I be concerned about the day of the week or the date?"

Additional gentle prodding revealed confusion about the identity of the president and the governor of the state in which she lived. She couldn't name more than five animals or four flowers over the span of thirty seconds. Most striking of all, she could not name or identify her grandchildren.

A MRI scan done later that day showed diffuse atrophy

of the brain with enlarged ventricles. A PET scan showed generalized reduction of activity, most marked in the parietal and temporal areas. Neuropsychological testing provided additional evidence that Paula suffered from dementia, most likely Alzheimer's disease.

• • •

Dementia is now the fourth-commonest cause of death in the United States. Its incidence increases dramatically with age: less than 1 percent of 40-year-olds, 6 percent of 65-year-olds, and somewhere between 20 and perhaps as many as 40 percent of 80-year-olds suffer from disabling forms of dementia. In short, aging itself is the greatest risk factor for falling prey to the illness. But these statistics do not imply that if you live long enough, you will inevitably come down with dementia. Dementia is a *disease*, not the inevitable consequence of aging. For this reason we should hold the same attitude toward dementia as we do toward any disease: learn as much as we can about it and take all available measures to prevent it.

While dementia may be the result of a host of different diseases, ranging from thyroid disorders to head injury, most dementia results from two illnesses: Alzheimer's and cerebrovascular disease (damage to the blood vessels supplying the brain). In recent years the incidence of cerebrovascular disease, the major cause of stroke, has dropped dramatically as a result of newer, more effective treatments for high blood pressure, obesity, and elevated cholesterol. Alzheimer's, however, remains largely untreatable, mostly because neuroscientists aren't certain of the nature and cause of the disorder. Indeed, it's fair to say that the illness remains almost as mysterious now as when it was first described.

In 1906 Alois Alzheimer, a German psychiatrist often depicted with a microscope in one hand and a cigar in the other, reported to his colleagues the strange case of a middle-aged woman who claimed people were trying to kill her. Initially she expressed strong feelings of jealousy toward her husband. A severe memory impairment followed soon thereafter. Eventually she lost the ability to make her way around her own house.

Her mental responses also were peculiar. When her doctors held up an object for her to identify, she chose correctly at first, but within moments forgot what she had seen. When writing, she repeated some letters, omitted others, and finally lapsed into confusion and absentmindedness. In her conversation she often expressed bewilderment, uttered the wrong words, or suddenly stopped talking altogether. When questioned, she did not remember the questions long enough to answer them and could not describe uses for common, everyday household utensils. She had no idea of time, place, or the identity of friends or relatives. Over several months she became increasingly confused, disoriented, and eventually delirious. Her illness relentlessly progressed over four and a half years, until she was totally bedridden and apathetic. She died in the institution at 51. After her death Alzheimer commented: "On the whole we are dealing with a peculiar, little-known disease process."

Within only a few years, other instances of the illness, eventually named after Alzheimer, frequently came to physicians' attention. As with the original case, many Alzheimer patients at first showed mild impairments of memory and lapses of attention. Early in the illness the symptoms were almost undetectable to anyone other than close relatives, who observed personality changes, frequent confusion, and the neglect of personal affairs. But eventually, as with Alzheimer's original patient, the affected patients showed hesitancy in their speech, inability to come up with the names for common objects, frequent disorientation, and memory loss for the identity of family and friends. In most instances, recent events were affected more than events from the distant past.

In the final stages, the Alzheimer patients stopped talking altogether and lapsed into vacuous silence. The patients could not or would not rise from their beds. Within weeks of these failures, death mercifully intervened in the form of pneumonia or other infections.

In the near century since Alzheimer's original description, neurologists have refined their understanding of the illness. They have discovered that it usually occurs in three distinct stages.

Stage one is marked by a subtle memory impairment, that, initially, other people may attribute to ordinary absentmindedness. But in fact it is not normal at any age to regularly forget what was eaten at lunch, one's companions at the meal, or the general topics of conversation while dining. The depth and degree of memory loss is important to remember and can save a lot of anxiety and worry among individuals of mature age. Every day, patients come to my office for neurological consultation because of anxiety over a perceived encroaching failure of memory. In the majority of instances, their concern is unfounded. Nor is my experience unusual. Fifty percent of 450 normal participants in an informational meeting on forgetfulness held in the Netherlands in 1995 indicated that they were worried about their memory and that they might be coming down with Alzheimer's disease.

The *second stage* of Alzheimer's disease is marked by language difficulties, such as failure to come up with the proper word and the employment of word substitutions; blunting of the emotions accompanied by lack of interest for people and events; and loss of the capacity for abstract thought and imagination.

In the *third and final stage*, all intellectual faculties are disrupted, and eventually, the personality totally destroyed. Relatives are no longer recognized, and if not prodded, the patient (the proper term, since at this stage full-time medical and custodial care is the rule) remains in bed until finally slipping into a vegetative state and eventual death.

Alzheimer compared the brain of his patient with the changes that often accompany normal aging. Overall her brain was dramatically reduced in size. Under the microscope, as many as 25 percent of the neurons had disappeared from certain parts of the cortex. Many of the remaining neurons contained dense, intricately tangled, thickened bundles of fibers. Overall, the brain of an Alzheimer patient is usually reduced in size and weight, averaging out to a loss of between 10 and 15 percent.

Until recently, neuroscientists disagreed about the relationship of normal changes in the healthy brain as it ages and Alzheimer's disease. One camp held that Alzheimer's is only an exaggerated form of aging that everyone is prone to if he lives long enough. In support of their contention, the proponents of this view pointed to the fact that many of the physical signs of Alzheimer's in the brain occur in normal brains. According to this theory, when they reach a certain critical threshold, they produce Alzheimer's disease.

Other scientists held to a qualitative theory: Alzheimer's is not simply the result of aging changes slowly accumulated over the years but, instead, is an aberration of the aging process. In support of this view was the discovery of genetic risk factors and the association of some forms of Alzheimer's disease with disturbances at specific chromosomal sites, notably chromosomes 14, 19, and 21.

In 1995 the quantitative and qualitative theories of Alzheimer's were put to the test. At autopsy, the brains of Alzheimer patients were compared to the brains of normally aged individuals who died of natural causes unrelated to brain disease. The brains of Alzheimer's patients differed from that of normal people in the patterns of nerve-cell loss in the hippocampal region, a center deep within the brain important in memory (a function that is gravely impaired in Alzheimer's disease). Such a finding has convinced neuroscientists that Alzheimer's is vastly different from normal aging.

"Alzheimer's disease is not the manifestation of accelerated aging, but the expression of a distinct pathological process," the investigators concluded.

Another proof that Alzheimer's isn't an inevitable consequence of aging comes from incidence data. As a person survives into the late 80s, his likelihood of falling prey to dementia actually *decreases*. This unexpected and hopeful finding is consistent with what scientists refer to as *critical age:* the decade of the greatest increase in illnesses.

Over a person's life span illnesses develop in clusters. The period of greatest health occurs during young adulthood. The most dangerous period comes between 70 and 79 years of age. In that decade physical and mental calamities of various degrees of seriousness afflict the older person approximately twice a month. Moreover, the critical age varies according to body systems. The critical age is the same for the brain as for the immune, endocrine, and cardiovascular systems (peaking between 70 and 79). Following this most dangerous decade, health relatively stabilizes. The rate of increase of senile dementia rises until about 80, when it begins to fall. At about age 95, the prevalence rate levels off to about half the rate found in octogenarians.

Thus senile dementia is *age related* (occurring within a specific age range) rather than *aging related* (a consequence of the aging process itself). Some geriatricians are even cautiously suggesting a true functional improvement at the end of the critical age. In terms of the brain, this means mental acuity can be retained and even enhanced in the 80s and beyond.

In February 1995 K. Ritchie, a neuropsychologist, reported in the *British Journal of Psychiatry* on the mental state of Mme. Jeanne Calment of Arles, the world's oldest woman, mentioned in Chapter One. Over a six-month period the psychologist put the woman with the longest authenticated life span in the history of the human species through a series of psychological tests aimed at discovering her mental capacities. Because of Mme. Calment's vision and hearing difficulties, some of the tests had to be improvised.

In order to appreciate how psychologists like Ritchie go about evaluating memory performance, imagine yourself participating in a commonly employed memory experiment. As you sit quietly in the laboratory, the psychologist reads a story aloud to you. After she is finished, you are asked a series of questions about the characters in the story. Ten minutes later the questions are repeated and the answers compared. The same thing is done two hours later and forty-eight hours later.

Studies of immediate recall (just after hearing the story) and delayed recall (at ten-minute, two-hour, and forty-eight-hour intervals) show a slow but steady decline in delayed recall in older people starting in the 50s! That last fact may surprise you, even shock and disappoint you. Actually, the falloff is very gradual and, with additional practice and re-hearsal, doesn't appreciably affect the ability to learn and remember. Of much greater importance is the fact that this slow incremental falloff in delayed recall is dramatically different from the performance of someone with Alzheimer's disease or other dementia. What is that difference?

Most people, when they take the memory test I just described, find it mildly amusing to be asked the same questions for the second time after an interval of only ten minutes. After all, anyone could answer the questions after a delay of only ten minutes—or so it would seem. And yet that brief ten-minute delay reliably distinguishes people with early Alzheimer's disease from normals of any age, from early adulthood on up.

Not only is memory failure the hallmark of early Alzheimer's disease, but it is very specific, with a lot of information disappearing early. For instance, if the subject is given a list of everyday words to be memorized for immediate and later recall, forgetting starts within ten minutes and is usually profound. Perhaps only one or two items will be remembered from a list of ten items—a dramatic and reliable distinguisher from the gradual decline in delayed recall observed to take place across the life span. The same thing happens if the subject is asked to copy a design figure of, say, a flower, and then, after a delay, to draw it strictly from memory. The resulting figure may consist of only a few squiggly lines.

As Marilyn Albert, a clinical neuropsychologist at Harvard University, puts it,

Alzheimer's disease patients demonstrate an increased rate of forgetting. They lose very early what they have learned. This differs significantly from normal aging, where the problem is a slower rate of learning with more time required to learn new information. But once information has been learned, the normal older person has as good a grip on it as a younger individual.

What accounts for the Alzheimer patient's loss of learned information within ten minutes? In the brain affected by Alzheimer's disease, as contrasted with the normally mature brain, characteristic plaques and tangles can be found in abundance in the hippocampus and medial temporal lobe. These abnormal accumulations are thought to cause a kind of short-circuiting that prevents the intake and consolidation of information. Research with monkeys shows a strong correlation between memory failures and hippocampal–medial temporal lobe disease. To confirm that this is the case in Alzheimer's disease, autopsy studies were performed on five older people with early but significant memory loss who died of unrelated causes (accidents, cancer, and so forth). The brains of all five of these people contained plaques and tangles in the hippocampus.

The frontal lobes are important also in developing strategies and coordinating ways of taking in information. In these tasks, the mature brain may be at a slight handicap. But once the information has been learned, as evidenced by hippocampal activation, it is as well retained in the old as in the young.

And this is exactly what was found by K. Ritchie in his test on Mme. Calment. At 118 years and 9 months, her verbal memory and language fluency were comparable to those of persons with the same education in their 80s and 90s. She showed no evidence of disturbance in frontal-lobe function. Ritchie concluded that Mme. Calment showed "no evidence of progressive neurological disease." Whatever mental deterioration Mme. Calment showed was the result of processes that had occurred two or more decades earlier: her brain was now functioning at a steady state. In short, Alzheimer's and other forms of dementia are *diseases*, not simply exaggerated degrees of aging. Thus the brain changes associated with dementing illness are not normal at any age. As a consequence, it's fair to say that fears about coming down with Alzheimer's are in most instances unjustified.

Despite a few promising candidates, there are no predictive tests that can guarantee early diagnosis for Alzheimer's disease. Just as there is no way of guaranteeing good physical and mental health, so there is no way of predicting who is destined to do poorly in the game of genetic roulette in which we are all players. And when you think about this situation, would you wish it any other way? Alzheimer's disease is currently incurable. Other than for reasons of estates and inheritance, what possible advantage would it be for you to know personally that in a few years time you will be afflicted with a mentally destructive and fatal brain disease? (Forget for the moment about the possible applications of such knowledge to the development of new treatments for other people.) Rather than seeking predictive tests, doesn't it seem to make much more sense, in the absence of evidence to the contrary, to assume you will remain free of dementia, and to learn as much as possible about brain functioning in the later years so you can make efforts to improve your brain?

• • •

While dementia evolves ever so slowly within the brain, mimicking the process of aging itself, one condition—*stroke*—occurs within the brain with the immediacy of a thunderclap. Indeed, Hippocrates wrote of stroke as *plēssō*, meaning "to be thunderstruck."

With or without an atmospheric connotation, stroke until recently was spoken about in terms implying inevitability, finality, and untreatability. Such fatalism is no longer justified, since now, for the first time in medical history, strokes can be prevented if symptoms are recognized and treated in time. Unfortunately, few people are aware of the signs of stroke. A questionnaire about stroke answered by a cross section of the population of suburban Virginia revealed that many of the respondents, most of them with college educations, could not enumerate any signs of stroke. Many of them did not even know for certain that a stroke involved the brain rather than the heart. Since stroke can be so devastating and yet is ultimately preventable and treatable when recognized, I have included this condition among those brain dysfunctions occurring primarily in the mature years that we should recognize.

A stroke results when a blood clot or ruptured artery serving the brain interrupts blood flow. The results of blood deprivation are particularly devastating to the brain, since it has a high metabolic rate. While the brain is only 2 percent of body weight, its metabolism accounts for 20 percent of the body's oxygen consumption, and it receives 15 percent of the blood pumped out of the heart.

Eighty percent of strokes result from the closure of a major artery supplying the brain. A *thrombotic stroke* occurs when blood flow is cut off by a clot resulting from the gradual buildup over many years of fatty deposits on the walls of an artery serving the brain. An *embolic stroke* results when a clot formed elsewhere in the body—often in the heart—breaks loose and lodges in an artery leading to the brain. The clot may be formed in the heart as a consequence of a recent heart attack; more commonly, it is precipitated by an abnormality in the heart's rhythm, such as atrial fibrillation, one of the risk factors for stroke.

The remaining 20 percent of strokes are *hemorrhagic*, which means they result from a rupture of an artery within (intracerebral) or around (subarachnoid) the brain. While intracerebral hemorrhages occur primarily in the later years as a result of high blood pressure, subarachnoid hemorrhages affect people at any age. That's because the hemorrhages stem from an inherited weakness in the wall of an artery, which, under the pressure exerted by the heart in moving blood throughout the body, gradually balloons outward into an aneurysm (an outpouching of the blood vessel wall). An aneurysm is a biological time bomb that can go off at any time during a person's lifetime and produce a devastating, often fatal outcome (as many as 50 percent of subarachnoid hemorrhages result in death).

A patient who is in the process of having a stroke may suffer a host of symptoms that may remit, in which case the episode is called a *transient ischemic attack*; or the symptoms may become permanent, that is, a *completed stroke*. Since transient ischemic attacks can often be treated, thus preventing the progression to completed stroke, everyone should be familiar with stroke symptoms. These include one-sided numbness; difficulty speaking or understanding other people's speech; sudden severe headache, dizziness, or loss of balance; visual difficulties such as blurred vision, decreased vision, or even the complete loss of sight in one eye. While these are the commonest initial symptoms of stroke, other symptoms may also occur depending on the stroke's location.

Right hemisphere:

- Left-sided weakness or paralysis
- Denial of or indifference to the paralysis
- Loss of sensation on the paralyzed side
- Loss of vision from both eyes in the left visual field
- Confusion, disorientation for time and location, emotional instability, dulled responsiveness, poor judgment, impaired ability for logical thought

Left hemisphere:

- Right-sided paralysis
- Loss of vision from both eyes in the right visual field
- Problems in speaking and understanding the communications of others
- Depression, slowness, impaired thinking, temporary confusion

Cerebellum:

- Loss of balance and equilibrium
- Dizziness, nausea, vomiting
- Flaccid, "rag doll" weakness of the arm and leg on the damaged side

Brain stem:

- Difficulty swallowing or pronouncing words
- Inability to walk a straight line, accompanied by severe vertigo, nausea, and vomiting
- Bilateral weakness or paralysis
- Unstable blood pressure and pulse leading to coma

As mentioned above, reversible stroke—like cutoffs in blood supply—can occur. A TIA, transient ischemic attack, is a temporary, localized, and self-reversing dysfunction caused by a reduction in the blood supply to the brain. The attack comes on usually within seconds and lasts anywhere from two to fifteen minutes, never more than twenty-four hours. Like strokes, TIAs also produce symptoms (patient complaints) along with signs (objective changes observed by the doctor). A TIA cannot always be reliably differentiated from a stroke, since the signs and symptoms may be the same. And since TIAs often serve as harbingers of strokes, they are true neurologic *emergencies* and demand prompt evaluation. Failures of diagnosis can be disastrous.

Five percent of people experiencing a TIA will suffer a

stroke within a month. Twelve percent will suffer one within a year; 25 percent will be felled at the end of three years.

The likelihood of both TIA and stroke can be reduced by lifestyle changes, such as exercising more, dieting, drinking less, and not smoking. Important medical measures include lowering the blood pressure in hypertensives, regulating cholesterol and blood sugar, thinning the blood with anticoagulants, and prescribing drugs that reduce the "stickiness" of platelets in the blood. The most readily available antiplatelet agent is aspirin. Taking just five grains a day reduces the risk of a TIA becoming a stroke by almost 25 percent. A prescription antiplatelet drug, ticlopidine, is 50 percent more effective during the first year after the TIA.

Another treatment, approved in June 1996 by the Food and Drug Administration, involves the administration of a clot-dissolving drug. When properly administered, tissue plasminogen activator (TPA) protects the brain from permanent injury. According to early clinical trials, a stroke victim given TPA is at least 33 percent more likely to recover or have minimal disability than someone not given the drug. But it must be given early—no later than three hours after the first signs of a stroke caused by a clot that blocks blood flow into the brain (an ischemic stroke). If given later than three hours, the drug itself can set off dangerous bleeding in the brain. And if the stroke is caused by hemorrhaging in the brain instead of ischemia, TPA will worsen the stoke and can even kill the patient.

The use of TPA is currently a topic of controversy among neurologists and other stroke specialists. Sometimes it can be very difficult even under the best circumstances to distinguish a stroke caused by ischemia from one resulting from a brain hemorrhage. In addition, the three-hour window of opportunity isn't always easily determined: the stroke may have begun several hours before the patient or anyone else noticed anything amiss. For these reasons, doctors and pharmaceutical manufacturers are looking in new directions, hoping to come up with a treatment less constraining and potentially dangerous than TPA.

•   •   •

In order to understand the new approaches to the prevention and treatment of stroke, it's necessary to discuss what happens in stroke at the level of brain cells.

Whatever the cause of stroke, the neurons in areas cut off from their regular blood supply die as a result of deprivation of oxygen, glucose, and other nourishing substances. And since neurons do not multiply or degenerate, the functions carried out by these cells either will be taken over by other neurons or will cease.

The degree of damage caused by the ischemia, the loss of blood supply, depends on how much and how long blood flow is reduced. Only rarely is the delivery of oxygen and nutrients cut off completely, thanks to the anatomical arrangement of blood vessels serving the brain.

As another saving feature, each of the main blood vessels gives off collaterals, which extend several millimeters beyond the termination of the main trunk. In addition, the blood vessels, unlike rigid pipes, are capable of regulating their diameters by autoregulation: dilating or constricting in order to increase blood-flow delivery in threatened areas.

Collaterals and autoregulation work together to minimize the effect of ischemia and limit neuronal cell death. As a result, the areas of damage caused by ischemia show a typical pattern: a central core of severe damage containing dead and dying neurons (the infarction), surrounded by a penumbra of damaged and dysfunctional neurons. The situation is similar to what can be observed after a bomb blast: a central core of dead or irreversibly injured people, surrounded by others with degrees of injury that vary with their distance from the blast.

The medical response to a stroke and a bomb blast are also similar. Physicians operate according to the method of triage: they assess the seriousness of damage and concentrate on the seriously injured whose treatment will make the difference between survival and death. Thus, neurons in the penumbra area are particularly important. If they can be saved, then the numbers of dead neurons and malfunctioning or destroyed neuronal circuits can be limited.

Experiments on animals suggest the existence of a window of opportunity during which treatment measures can rescue the neurons in the penumbra. But treatment must come relatively soon because, as experimenters have learned from animal research, neurons may cease to function and literally begin to dissolve anytime between one hour and four hours after the onset of ischemia. Since no one knows for certain just when the point of irreversibility is reached in humans, current research on new drugs focuses on delivering treatment during the first six hours after a stroke.

In order to understand the new approach to stroke treatment, it's necessary to delve down even deeper than the neuron, leaving the world of directly observable phenomena and entering into the chemical and molecular world. There, neuroscientists have discovered some surprising and even paradoxical events.

For instance, the most common brain neurotransmitter, *glutamate*, can function as a kind of double agent, depending on its concentration and location. In normal concentrations and under normal conditions, glutamate is the principal excitatory neurotransmitter within the brain. Its reactions with special receptors on the nerve cell membranes are responsible for many brain activities, such as memory, thinking, movement, and sensation. At higher doses, however, glutamate is a killer of brain cells. The excess glutamate overstimulates the glutamate receptors on the cell membrane in a process neuroscientists term *excitotoxic injury*. The paradox I mentioned a moment ago is this: How and why did the brain evolve with a vulnerability to damage and death resulting from the action of its most ubiquitous messenger? Since neuroscientists found such a situation inexplicable, they began several years ago to carry out the necessary experiments leading up to the formulation of the excitotoxic theory.

In essence, the excitotoxic theory is based on the finding that, when damaged or deprived of blood, injured brain cells spew out their contents of glutamate onto the doorstep of the other neurons in the immediate area. This wave of extracellular glutamate rises to flood levels and overwhelms the glutamate receptors on the membranes of the nearby cells. Thus, the wave of injury is self-propagating: injured neurons release large amounts of glutamate, which in turn wash over and injure other cells, which then release their glutamate into the pool and thus injure their neighbors . . . and so on.

The exact mechanism underlying the injury caused by glu-

tamate remains an area of intense investigation. Both immediate and delayed forms of injury occur. The initial sign of
injury, which takes place in the first few minutes, consists of
a swelling of the neuron. This is due to disruption of the
stability of the nerve-cell membrane. As a result, an excess of
sodium and chloride ions, together with water, then enters the
cell. Though serious, this neuronal swelling and its immediate
consequences are potentially reversible. More serious, and not
as reversible, are the delayed effects of injury. These depend
on the action of glutamate on its special receptors on the
nerve-cell membrane.

Every nerve-cell membrane has at least four different receptors for glutamate. For the sake of simplicity these receptors are named after laboratory chemicals that can activate
them. Thus, the NMDA-glutamate receptor is named after
the activating chemical N-methyl-D-aspartate. Overstimula-
tion of the NMDA receptor can produce nerve-cell damage
because the receptor allows free passage, across the membrane, of calcium, an element that in excess is highly toxic to
neurons.

The sequence goes like this: The NMDA receptor in the
presence of excess glutamate permits a large and rapid inflow
of calcium into the neuron. Like glutamate, calcium is helpful
and necessary within limits, but in excess turns into a killer.

If too much calcium enters the cell through the NMDA
receptor, it triggers a cascade of chemical processes within the
cell. The most damaging is the activation of enzymes that act
like acid poured onto the surface of a computer chip. They
deform and destroy the internal structure of the neuron, halt
all transport within the cell, and further increase the perme-
ability of the cell's outer membrane.

Not all parts of the brain are equally susceptible. Neurons in different parts of the brain vary in their vulnerability to glutamate-mediated damage, according to their location and the density of NMDA-glutamate receptors on their cell membranes. The hippocampus, important in memory, and the cortex, the mediator of thinking, are among the most vulnerable. Since the cell membranes of these brain areas are densely packed with NMDA-glutamate receptors, even short episodes of ischemia and resulting glutamate-induced damage can lead to memory loss and deficits in thinking.

A new and revolutionary treatment approach to stroke involves the administration of drugs that oppose the action of the NMDA receptor by blocking its actions. As mentioned a moment ago, the receptor ordinarily acts as a gating mechanism for the ions sodium, potassium, and calcium. The operation is similar to the opening of a dike: when the gate is opened, each ion can flow in and out of the neuron through its respective channel. Thus, when glutamate attaches to its NMDA receptor, the channel opens and calcium enters. If excess glutamate latches onto the NMDA receptor and overstimulates it, excess calcium rushes inward to wreak the damage mentioned a moment ago.

An open-channel blocking drug works by binding to one or more of the receptor's ion channels, thereby impeding the passage of those ions into the neuron. Since calcium is the ultimate villain in the glutamate-induced damage, current treatments involve calcium channel blockers. In animals these blockers can reduce the amount of stroke damage by as much as 88 percent.

Unfortunately open-channel blockers other than for calcium can't be used in humans because some of them induce psychotic reactions. The street drug PCP (angel dust) is an excellent open-channel blocker that reduces the damage of stroke, but at the cost of bodywide anesthesia, impaired coordination, slurred speech, and frightening out-of-body experiences. These symptoms and signs can rapidly progress to bizarre behavior, combativeness, coma, and even death. Dextrorphan, a breakdown product of the cough suppressant dextromethorphan, is another potential open-channel blocker lacking the serious liabilities of PCP. It is currently under study for treatment of stroke.

Another approach to stroke is to block the glutamate site on its NMDA receptor. If glutamate never makes contact with its receptor, then the calcium channel is not opened, calcium influx is stopped, and nerve-cell damage is prevented. Although this seems like a reasonable approach—it has worked well in animals—blocking the glutamate receptor entirely is an example of too much of a good thing: total glutamate-receptor blockade disturbs the normal functions mediated by glutamate in proper amounts. Thus thinking, memory, and other mental processes ordinarily mediated by glutamate's excitatory transmission may be interfered with. Glutamate is an enemy, recall, only when present in excessive amounts. At normal levels it is a friend that enables the brain to operate most efficiently.

One other promising approach to stroke therapy involves blockade of another site on the NMDA receptor. Antagonists called glystasins are used to target the binding site of the neurotransmitter glycine. These agents are safer and do not interfere as much with normal thinking or induce psychotic states.

Within the next year or two successful drug treatments will

be available that will reverse many of the destructive effects of stroke. Nihilism and acceptance will be replaced with optimism and aggressive treatments. *Three new perceptions* are particularly important:

- Stroke is largely preventable via health and lifestyle changes.

- Stroke requires emergency treatment at the earliest possible moment.

- Stroke is a "brain attack," according to the National Stroke Association, and should be handled with the same urgency as a heart attack.

Each of these new perceptions emerges from the experiments on excitotoxic damage. They also carry important and practical treatment implications. For instance, since the glutamate–calcium–nerve-cell injury sequence happens so quickly, it is necessary to begin treatment of the stroke patient as quickly as possible. Thus ambulance and emergency medical personnel are currently undergoing reeducation on the need to respond to calls describing potential strokes with the same sense of urgency they would display toward any life-threatening, reversible emergency.

Within the next five years it is expected that brain-protective drugs will be available that can be administered in the ambulance on the way to the hospital. On arrival, the patient will then be rapidly evaluated as to whether he or she is likely to benefit from additional drugs capable of breaking up clots in the arteries leading to the brain.

Even in completed strokes, the chances for cognitive recovery are greater than commonly believed. Most neurologists have been taught that improvement following a stroke takes place in the first three months, and that little additional improvement can be expected after that. But a recent study from the Neurological Institute of Columbia-Presbyterian Medical Center in New York shows that stroke patients can continue to show positive gains for up to two years after stroke. In the study, patient scores on verbal and nonverbal memory tests were significantly better six months following stroke than they had been three months after stroke. Stroke can now be controlled and its destructive effects minimized by early diagnosis, prompt treatment, and extended neurorehabilitation.

7

THE BRAIN'S

CHEMICAL MESSENGERS

*I*n understanding Alzheimer's and other dementias, as well as other problems affecting the mature brain, we need to learn about brain
chemicals and how they function at the microscopic level of
the neuron. Treatments for these disorders, in turn, involve
chemicals and drugs that equalize some of the imbalances
between the younger and the more mature brain.

The brain consists of 100 billion special cells called neurons.
Unlike other cells in the human body, neurons, once formed, do
not as a general rule divide. Also in contrast to other body cells,
neurons do not deteriorate with usage, or wear and tear, but
perform better and more efficiently with increased activity.
This quality of the neuron is unique and has great practical implications. Indeed the injunction that we "use it or lose it" is
based on this unique quality of human neurons.

Neurons communicate with one another in tag-team fashion by means of electrical and chemical signals. Communication between two neurons starts with an electrical signal, the nerve impulse, that travels along the length of a specialized extension of the neuron called the *axon* until it reaches the terminal point of the axon. Here the nerve signal is converted from an electrical signal to chemical transmission. Calcium enters the axon and stimulates the release of a chemical messenger, the neurotransmitter, which slips across a tiny junction, the *synapse,* separating the two neurons. After completing the passage across the synapse and arriving at the receiving dock (dendrite), the neurotransmitter latches onto a specialized receptor that is " turned on," thereby generating a nerve impulse in that cell.

The outer two to five millimeters of the cerebral hemispheres, the cerebral cortex, are highly developed in humans and consist of several billion neurons. Moreover, these neurons are richly interconnected. A single cell may be linked with several thousand others. This dense interconnectivity is responsible for the richness of human thought. Damage to or loss of any of these neurons results in mental failure of one sort or another, depending on the location of the loss or damage.

Healthy brain activity in the later years depends on the healthy functioning of fifty or more neurotransmitters, the chemicals that convey information throughout the brain from one neuron to another. If you want to understand how the brain changes functionally with aging, therefore, it's necessary to look at neurochemistry: how the brain functions chemically, the chemical changes that accompany aging, and what can be done to reverse the harmful effect of these chemical changes.

While anatomy is comparable to the architectural layout of a building, neurochemistry is more dynamic and corresponds to the electrical wiring that turns on the lights and runs the appliances. As mentioned, information processing in the brain involves the sending and processing of nerve-cell signals via the action of neurotransmitters and receptors for these neurotransmitters. In general, the main neurotransmitters in the brain decrease with aging; a similar reduction takes place in the numbers of receptors for the neurotransmitters. And these reductions carry consequences.

For instance, groups of neurons called nuclei, hidden deep within the brain, manufacture a neurotransmitter called dopamine, which is important in the smooth execution of movements. If these nuclei are damaged or the amount of dopamine produced by its neurons diminished, the affected person develops a tremor and other indications of Parkinson's disease.

Another example. Dopamine receptors in the frontal cortex decrease in density as a person ages. As a result, certain frontal-lobe dysfunctions, particularly those involving working memory (keeping prior information in mind while working with new information), are selectively affected. This loss of frontal dopamine receptors is at least a partial explanation for the decline in some older people's ability to carry out, simultaneously, multiple mental processes. The good news is that something can be done to reverse these chemical changes. For instance, the loss of a neurotransmitter can be compensated for by the judicious use of medications. As an example, thousands of Parkinson's disease patients are bene-fiting from replacement of the neurotransmitter dopamine. As Parkinson's disease worsens, increasing numbers of neurons die off in a particular nucleus in the brain stem called the substantia nigra (the name derives from the structure's black appearance to the naked eye). But many of the signs of Parkinson's disease can be reversed when the patient is given dopa, a precursor molecule that is transformed in the brain into dopamine. Dopa substitutes for the missing natural neurotransmitter. Such replacement therapy is not possible if receptor cells for dopamine are destroyed by the illness. With no receptor to latch onto, the dopamine floats around like a yacht without a slip. Under such conditions all the dopamine in the world cannot make the patients better.

Think of an old lawn mower that has gone rusty and clogged due to disuse and the passage of time. Some oil and servicing can restore things to near normal. But the addition of oil and simple servicing can do nothing if the blades are twisted and parts are missing. Neurotransmitter replacement can activate brain areas that have fallen into disuse but still remain viable.

Increasing knowledge about neurotransmitters and their receptors has enabled neuroscientists to come up with drugs capable of reversing the effects of several age-related disturbances.

For instance, the mature brain is especially susceptible to several mental and emotional disorders, including depression, sleep problems, anxiety, alcoholism, and deficiencies in sexual desire and performance. Neuroscientists are confident that these disorders stem from brain disturbances, but the correlations are less well established than for Alzheimer's disease.

Chemical and behavioral treatments for each of these conditions are now available—thus narrowing the gap between the performance of the older and younger brain.

Foremost among the mental health problems of the mature brain are depression and related suicide. But in order to put these into perspective it's necessary to say a few words about the most pervasive emotional dysfunction in the later years: loneliness.

• • •

Loneliness is the greatest challenge of the mature years. The loss of relatives and friends leads to an inevitable loneliness in the survivors. Indeed, when it comes to accepting and dealing with loss, each day presents its own challenge. Simple everyday actions can serve as a trigger for the ensuing feeling of loneliness: perhaps reading in the morning paper the obituary of a former acquaintance or coworker. One half of all men and women 90 years or older report persistent feelings of loneliness. Among the bereaved who have lost a husband or wife within the past year, loneliness is judged even more difficult to endure than low mood, fatigue, anxiety, and problems in memory and thinking.

Indeed, the intensity of loneliness in later life is sufficiently severe that fully one half of otherwise normal bereaved widows and widowers admit to experiencing illusions or hallucinations of the deceased. About a third tell their doctors they have seen, heard, or even talked to their dead husband or wife. Such hallucinations, along with illusions, occur most frequently in persons whose marriage was happy.

But despite such poignant experiences—which do not in any way indicate mental illness—older people actually adjust better on the whole to loss and suffering than their younger counterparts. As a rule they undergo less depression and fewer anxiety symptoms. Indeed, the oldest widows and widowers show the most consistent improvement in their levels of distress and loneliness over time. They do this by relying on friends rather than family for emotional support.

Numerous studies show that an older person's morale usually depends more on how often she gets together with her friends than on how often she sees her children. In a fascinating research project ninety-two retired adults (fifty-two women and forty men) between the ages of 55 and 88 were outfitted with beepers, which they carried with them everywhere for a week. At irregular moments they were paged. At the instant of paging, the participants in the experiment wrote down whom and what they were thinking about. Overwhelmingly they reported mental involvement with friends rather than family.

Successfully aging persons, it seems, get more excited and emotionally aroused with friends than with family members, including their spouses if they are still alive. I suspect that's because friends provide a respite from the monotony of daily routine and an outlet for confiding thoughts and feelings. People who can talk about their worries and fears with friends are better qualified to deal with the challenges of aging. They also live longer—perhaps one of the reasons why women, who are more comfortable with self-disclosure, tend to live longer than men, whose behavior often conforms to the cultural stereotype that equates self-disclosure with weakness.

When it comes to coping with loss, older people are not at all at a disadvantage. By this time in their lives, most older people have become more philosophical: they are accustomed to change, and accepting of the loss that frequently accompanies it. This resiliency should be kept in mind before concluding that depression is a normal response to aging.

• • •

A typical example of depression in the later years is a woman well known to me, whom, to protect her identity, I will call Arlene.

While visiting her son at the beach, Arlene, an 80-year-old widow known for her energy and upbeat approach to life, slipped on the stairs and suffered a broken wrist along with a chip fracture of the hip. No treatment was needed other than setting the bones in the wrist and taking pain pills to control the hip pain.

Within two weeks of the fall, Arlene was no longer upbeat but moody and despondent. She complained of many different symptoms, which her doctor could not explain. She told her daughter-in-law that she didn't expect to recover and would probably die of her injuries. Since this dire appraisal sounded unreasonable, the daughter-in-law told her husband, Arlene's son, about it. A physician himself, the son recognized that his mother was depressed and asked one of his colleagues to look in on her.

Arlene's lack of energy, fatalistic mood, sleep disturbance, and recurrent unreasonable thoughts of death confirmed the diagnosis of depression. A low dose of the antidepressant drug Zoloft was started.

The earliest generation of antidepressants, dating from the 1960s, worked by modifying several neurotransmitters rather than a single neurotransmitter. Unfortunately, this often resulted in severely limiting side effects, such as dangerous drops in blood pressure, which often resulted in falls and broken bones.

Further progress in psychopharmacology (the art and science of administering mood-altering drugs) led to the development of a new class of antidepressants, with fewer limiting and serious side effects. Among the drugs most helpful to later-life depression is a class of drugs aimed at influencing only a single neurotransmitter.

Particularly effective in the older person are the newer antidepressants, the *serotonin reuptake blockers*, such as Prozac, Paxil, and Zoloft.

Zoloft is one of a class of antidepressant drugs that work by modifying the influence within the synapse of the neurotransmitter serotonin. Ordinarily, after its release from one neuron (referred to as the presynaptic neuron), serotonin travels across the synaptic cleft and latches onto its own receptor on the postsynaptic neuron. This interaction of receptor and its specific receptor, sometimes likened to the action of a key fitting into a lock, sets off a series of events within the postsynaptic neuron. Milliseconds after locking on to its specific receptor, the serotonin is released and is taken up once again by the first neuron. Information is transferred within the brain by means of this action of neurotransmitters upon their receptors.

For reasons no one understands, depression can often be relieved by drugs like Zoloft, which interfere with the reuptake of serotonin from the synapse. Such drugs are referred to as serotonin-reuptake-inhibiting drugs, or SSRIs. But despite the lack of an explanation for the success of serotonin-reuptake-inhibiting drugs, one thing is clear: depression is a chemical disease that can be treated and cured by chemical agents. This insight represents a major conceptual break-through.

Until the 1960s, depression was considered the result of everything from demonic possession to the harboring of mur-derous feelings toward one's parents. But all of that changed with the development of antidepressants like Zoloft and other classes of mood-elevating agents. A chemical treatment speaks convincingly of a chemical cause—not the myriad of psychological and social "explanations" popular in the past.

While depression may affect people of any age, it takes its greatest toll among older persons. This is the result of age-related alterations in the numbers of neurotransmitters and their interactions. As a result of these neurotransmitter changes, the older brain is more susceptible to depression. The positive aspect of this aging process is that corrective, even prophylactic, measures can be started early enough in a depression to make a real difference in outcome. As a general rule, the earlier the depression is recognized, the quicker and more effectively will the antidepressant drug work. In Ar-lene's case the depression was moderately advanced when her son recognized the signs of the illness and obtained help for her.

None of the findings on the chemical nature of depression imply that the illness should be managed simply by the pre-scription of an antidepressant. Arlene also benefited from speaking to her psychiatrist. But rather than talking about unresolved lifelong conflicts that had little relevance to a woman of 80 who, prior to her fall and subsequent depres-sion, was aging successfully, Arlene's psychiatrist concen-trated instead on explaining to his patient that she was depressed. He assured her that as the antidepressant took effect, many if not all of her physical complaints would disappear.

Within two weeks after starting Zoloft, Arlene was calling her friends again, planning a trip with her son and daughter-in-law, and generally responding to the people and events with her former enthusiasm. Six months later she continues to do well, and her antidepressant dose is now reduced to only one pill taken just before bed.

Arlene's illness and its response to antidepressant medication is typical of the depression that occurs in the older person. As with Arlene, it's not unusual for the depressed person, and often the doctor as well, to fail initially to recognize the symptoms. This is because the sadness is often intermixed with widespread and multiple physical complaints. This absence of recognition of the depression leads to more visits to the doctor, with unexplainable symptoms, poor cooperation with treatment (often the medication offered not only doesn't help but leads to troublesome side effects), and increased risk for death by suicide.

Age-associated variations occur in the symptoms of depression. Younger people, when depressed, tend to eat and sleep more, gain weight, lose interest in sex, suffer from headaches, and experience their worst depressive feelings late in the afternoon or in the early evening. Older depressed people, in contrast, complain of insomnia, endure periodic attacks of agitation, and often suffer their worst depressions in the early morning.

Rarely does a person suffering from late-onset depression complain only of low mood. Typically, he or she experiences and describes loss of appetite, lack of energy, and decreased interest in and enjoyment of normal events and activities.

As with Arlene, most elderly depressives live alone; they tend to be widowed, divorced, or never married. Also characteristic are the recent loss of a loved one (Arlene's husband had died three years earlier, after a long illness) and heightened physical or emotional stresses (such as Arlene's fall and the ensuing need for her to remain inactive during the six weeks required for bone healing).

Another factor complicating the recognition of depression in the older person is the absence of a history of depression earlier in life. As with Arlene, the elderly depressive often does not understand what's happening because he or she has no previous depressive experience to compare it to.

After treating many cases of depression in older patients over the past twenty years and reading the burgeoning literature on the brain changes associated with aging, I am convinced that the majority of depressions occurring in later adult life are purely chemical in nature. In other words, older people are not depressed primarily because of psychological factors, fears of impending mortality, physical limitations, or impaired health—although each of these certainly contributes—but because of deficits and alterations of the neurotransmitters within the brain. As we age, some of the neurotransmitters are produced in lesser amounts, allowing others to exert a disproportionate influence on brain activity. Unfortunately, this influence may be an unfavorable one, leading to depression.

The good news is that, as with Arlene, low doses of antidepressants can restore the balance of neurotransmitters to normal. Several kinds of antidepressant drugs are available. All improve mood, restore restful sleep, stimulate the appetite, and heighten energy levels.

Finally, in those instances where antidepressants don't work, many older depressed patients respond extremely well to electroconvulsive therapy (ECT), a treatment often unrealistically feared and disparaged as "shock therapy." Actually, the electrical currents employed in ECT are extremely low and present no immediate or long-term harm to the patient.

Neuroscientists believe ECT works also by modifying neurotransmitters. It releases dopamine and other neurotransmitters in key brain areas. The action is specific; ECT is no longer considered as acting simply like a global reset button in the brain.

Since late-life depressions are chemically based and chemically responsive, it is critically important that older persons be placed on these agents at the first sign of depression. In most instances they will respond to low doses. In fact, the effective dose is usually a tenth or even a twentieth of what is required in younger adults. Such minuscule doses work so well because, I believe, the depression is the outward expression of a deficiency state. Age-related increases or decreases in the influence of neurotransmitters can be thought of as similar to a deficiency in a vitamin or other necessary nutrient. The situation is similar to the relative deficiency of the "juice machine" generators responsible for energy, mentioned earlier.

As a practical matter, every reader of this book should become familiar with the symptoms and signs of depression. At the first hint of depression in oneself or others, help should be sought from a neuropsychiatrist trained to consider emotional and mental disturbances as expressions of brain dysfunction.

Depression can usefully be divided into major and minor forms. Presumably, although not every expert is in agreement here, a major depression evolves from a minor one. Minor depression is less intense and involves a smaller number of depressive symptoms. Minor depressions, the usual predictors for major depressions, increase in frequency with age in what statisticians refer to as a *curvilinear* pattern. An increase occurs in people in their 30s, followed by a decrease in middle age and a very deep increase in people in their 80s or older.

According to the *Diagnostic and Statistical Manual of Mental Disorders IV*, the diagnosis of a major depressive episode requires that five (or more) of the following symptoms be present during the same two-week period and that they represent a change from previous functioning. At least one of the symptoms must be either *depressed mood* or *loss of interest or pleasure.*

- Depressed mood most of the day, nearly every day, as indicated by either subjective reports (feels sad or empty) or observation made by others (appears tearful)

- Markedly diminished interest or pleasure in all, or almost all, activities most of the day, nearly every day (as indicated by either subjective account or observation made by others)

- Significant weight loss when not dieting or weight gain (for example, a change of more than 5 percent of body weight in a month), or decrease or increase in appetite nearly every day

- Insomnia or hypersomnia nearly every day

- Psychomotor agitation or retardation nearly every day (observable to others, not merely subjective feelings of restlessness or being slowed down)

- Fatigue or loss of energy nearly every day

- Feelings of worthlessness or excessive or inappropriate guilt (which may be delusional) nearly every day (not merely self-reproach or guilt about being sick)

- Diminished ability to think or concentrate, or indecisiveness, nearly every day (either by subjective account or observed by others)

- Recurrent thoughts of death (not just fear of dying), recurrent suicidal ideation without a specific plan for committing suicide

In an older person, as contrasted to a younger one, depression is more often associated with some form of physical illness, as occurred with Arlene. Indeed, depression should serve as an early warning sign that something is physically amiss. In Arlene's case her physical problem was easy to diagnose. In most instances the diagnosis may prove more elusive.

In one study carried out among older people in Singapore, physical disorders were present in every depressed patient under study, with an average of almost two disorders per patient. Cardiovascular disease (poor blood flow to the heart and/or hypertension) and diabetes were present in seven-eighths of the patients. The depressive symptoms consisted of low mood and physical complaints that eluded all attempts at diagnosis. The depression was serious enough to cause sleep disturbances in 40 percent, while 30 percent were depressed enough to attempt suicide.

In addition, the depression in older people may be marked by *melancholic features*, which include depression regularly worse in the morning; early morning awakening (at least two hours before the desired time of awakening); lack of reactivity to usually pleasurable stimuli (the depressed person does not feel better, even temporarily, when something good happens).

In terms of incidence, somewhere between 15 percent and 25 percent of people over 65 are seriously depressed. As well as by medical conditions like heart disease or cancer, their depression can be triggered by the loss of a spouse or loved one (even a pet) or a decrease in vision, hearing, or ability to walk—conditions that disproportionately affect older people.

In nursing homes, the figures for depression are even higher: two-thirds of all residents show at least one symptom of depression. No doubt the isolating conditions in these homes contribute heavily to the high depression rate. Separated from family, friends, and familiar surroundings, the nursing home dweller can easily suffer a loss of context, essentially an identity problem. Moreover, the older person's difficulties are seldom recognized, even by medical personnel, as the signs of a depression. Only 4 percent of nursing home residents are ever offered mental health services. Nor are the workers in nursing homes given much in the way of mental health training. As a result, depression routinely goes unrecognized and untreated.

Why is depression so common in the later years? A youth-obsessed culture like our own puts the blame on the mere fact of getting older: "Nobody wants to get old, and therefore

people are understandably depressed in their mature years"
is how it's often expressed. So prevalent is this mistaken belief
that even older people themselves often believe it. A 1996
poll found that more than one-half of Americans over the age
of 75 think that late-life depression is a normal response to
aging. But this notion is simply mistaken. The vast majority
of older people are not depressed, even some of those afflicted
with seriously limiting diseases. It is terribly important, therefore, to recognize depression during the mature years as an
illness and to provide prompt treatment for the disorder when
it arises.

Over the next decade, as we learn more about the mature
brain, I expect that low doses of antidepressants will be prescribed to many older people on a routine basis for the mild
depression that can result from neurotransmitter changes.
These medicines will not function as "uppers," or stimulants,
although they will increase energy and well-being, but as
restoratives of the neurotransmitter balance that character-
izes good mental health.

Recognition of depression in the elderly is literally a matter
of life and death. That's because older people in the United
States have a higher suicide rate than any other age group.
Between 1980 and 1992 nearly seventy-five thousand Americans age 65 or older killed themselves—19 percent of the
almost four hundred thousand suicides in those twelve years.
The rate for white men over 65 is more than double that of
adolescents. Americans 65 or older account for about 13 percent of the population but account for about 20 percent of all
suicides. Men commit 81 percent of the suicides (38.4 suicides per 100,000), while the rate for women is about 6 per
100,000.

One of the reasons for this imbalance of the old over the young when it comes to successful suicides is that older people use more violent and lethal means to end their lives. Indeed, older people do not as a rule make suicide gestures: every attempt is literally deadly serious. Guns are the most common method chosen for suicide, used by 74 percent of men and 31 percent of women.

Older people take their lives for different reasons than their younger counterparts. Younger people commit suicide because of serious family disagreements, lack of money, unemployment, and loss of home and/or family. Alcohol plays a big role in many suicides of youth and early middle age. Older people kill themselves because they are in poor health, socially isolated, alcohol dependent, or unable to adjust to retirement.

Are such suicides a "rational" response to uncontrollable situations? Organizations such as the Hemlock Society make such claims; in 1991 a suicide manual became a best-seller. But there are several good reasons for skepticism. Suicides employing the most violent means often cannot be correlated with anything happening in the victim's life just prior to the suicide; the deaths remain totally inexplicable to the survivors. My use of the word "victim" in that last sentence was not casually chosen. About 90 percent of older people who attempt or complete suicide suffer from some mental disorder or disturbance. Often the person who elects suicide is in the throes of grief, alcoholism, or loneliness.

But most frequently the person is depressed. Indeed, the vast majority of suicides are caused by depression. In addition, the preceding depression has usually not been recognized or treated. Those depressives lucky enough to have been identified often receive inadequate treatment; hence their relapse and subsequent suicide.

Suicide in the older person can often be prevented via the successful treatment of the underlying depression. As with Arlene, older depressed people can respond dramatically to a combination of drugs and a few orienting psychotherapy sessions that concentrate on the present. Psychotherapy works by generally enhancing coping skills, improving attitudes, and clarifying the deleterious effects of negative thought patterns and behavior. The most successful forms of psychotherapy in the older depressed person do not involve detailed probing into events of the past. That's because, as mentioned earlier, depression later in life is almost always associated with some physical, often neurological impairment (Arlene's MRI showed some evidence of a small stroke that had occurred sometime in the past).

Of course, drugs and psychotherapy are not the only means of prevention of late-life suicide and depression. Physical fitness plays a role in maintaining a positive, healthy mental outlook. "Staying socially connected and maintaining a high level of physical fitness is helpful," according to Charles F. Reynolds III, a professor of psychiatry at the University of Pittsburgh School of Medicine.

In a classic paper, "In the eye of the beholder: Views of psychological well-being among middle-aged and older adults," geriatric psychologist C. D. Ryff compared the responses of people of average age 52 with others of average age 73 when asked to describe "good" or "poor" adjustment for their age and to define "successful aging." Both groups spoke mostly in terms of relationships with other people. Getting along with and caring about others, it turns out, is far more important to psychological adjustment than the self-directed, assertive, "me first" orientations too frequently emphasized by some psychologists (self-confidence, achievement, self-acceptance, self-knowledge, and so forth).

With age and the inevitable loss of friends and relatives, increased efforts must be made to make new friends. My patient Arlene includes making new friends as one part of her prescription for maintaining good health: "Take an interest in other people and keep abreast of what is going on in their world. Exercise and go out for a walk every day, even if it's only for a little while. Try to have a number of friends, and try to be a good friend. Keep up your physical appearance, because if you feel you look good, you will have more confidence and will be better able to meet people and make new friends."

• • •

In addition to depression, the mature brain is susceptible to difficulties with anxiety, sleep, and alcohol use. Although usually less disabling and impairing than depression, each of these difficulties can wreak havoc on physical and mental health. Fortunately, they can also be prevented or cured.

As a general rule, the nature and causes of anxiety change according to one's age and situation. As a person grows older, anxiety arises more from life circumstances than psychological factors. In the later years, the classical phobias and compulsions of young and mid-adulthood are encountered much less frequently and rarely break out for the first time. Instead, the older person often expresses increased and sometimes perfectly appropriate anxiety over personal safety. Indeed, statistics suggest that anxiety on this topic is not unreasonable.

The older you are, the more likely you are to be the victim of crime committed against you by strangers. You are also more likely to be exploited and exposed to scams and other ploys directed at securing money and property. Anxiety about such threats makes it difficult for an older person to experience a sense of control and mastery over his or her life. More time is spent alone in the house or apartment. This, of course, only adds to the resulting isolation and worsens the loneliness.

Anxiety in the later years frequently stems also from physical causes that often can be easily modified. Caffeine is a common villain, particularly when multiple cups of coffee or tea are combined with over-the-counter cold or allergy remedies, many of which contain caffeine, ephedrine, and other stimulating drugs. Each of these substances produces changes in neurotransmitters and other chemicals within the brain. Neurochemical changes resulting in anxiety can also occur with the withdrawal from sedative and tranquilizing drugs. As the level of the calming drug begins to drop, a "rebound" of anxiety may ensue. Among physical conditions, abnormal heart rhythms and thyroid disease may induce anxiety.

Treatment of anxiety often requires no more than reassurance and support. Many an anxious older person has been calmed by displays of concern and attention on the part of relatives and friends. When this doesn't work, antianxiety drugs like Valium can prove helpful. But the use of antianxiety drugs in the mature years can be tricky, since as we grow older we are more likely to develop side effects with standard adult doses of medications. More important, anxiety may actually be the outward manifestation of another illness, especially depression. Although the symptoms may sound to the doctor like anxiety, the underlying illness is often depression. Nor is this distinction a trivial one, since antianxiety drugs worsen depression, thus resulting in a cruel paradox for patient and doctor alike: the depression is temporarily relieved by the tranquilizer at the expense of a deepening and potentially fatal depression.

• • •

Disturbed sleep is the third common mental health problem to which the older brain is especially susceptible. The most common sleep difficulties are insomnia (difficulty falling or remaining asleep), breathing problems such as sleep apnea (momentary stoppages of breathing), episodic leg movements that lead to awakening ("restless legs"), and dysfunctional sleep patterns. One common sleep disturbance consists of early-evening drowsiness and early sleep onset, followed by middle-of-the-night awakening in the fully alert state.

As many as half of people over 65 suffer from chronic sleep disturbance. While a 20-year-old falls asleep within eight minutes of hitting the pillow, it takes an 80-year-old eighteen minutes to fall asleep. The quality of sleep also varies. Immediately upon falling asleep, the young person spends half an hour or longer in deep, restful sleep. His older counterpart, in contrast, may spend no time or at best only a few minutes in deep sleep.

Electroencephalogram (EEG) recordings show characteristic variations that explain the subjective sleep complaints. The duration of deep sleep is decreased from the 20 to 25 percent that occurs in a 30-year-old to 5 to 10 percent after age 70.

These natural changes in sleeping pattern increase with age

and render the older person's sleep more vulnerable to events happening in the immediate environment: traffic sounds, light entering under the doorjamb, the rhythm of rain on the roof. But frequent awakening from sleep can occur even in the absence of any stimulation; older people awaken frequently even in the sensory isolation of sleep laboratories, where sights and sounds can be artificially excluded. Thus the disturbed sleep pattern with its frequent awakenings does not represent an illness but rather a reduction in the quality of sleep brought about by the unavoidable process of aging. *Sixty Minutes* commentator Andy Rooney captured the dilemma in an essay, "Rock-a Bye Baby Is Long Gone." Rooney writes:

Sleep has changed for me. It is no longer a carefree, unconscious time for physical and mental resupply. I don't enjoy sleep the way I used to . . . maybe my memory is bad, but it seems as though I used to get into bed, go to sleep and wake up eight hours later. If that was true, those days are gone forever. Now I go to sleep quickly but no longer sleep for long.

Normal difficulties in falling and remaining asleep must be distinguished from disorders resulting from mental or physical illness. Signs of a pathologic sleep disturbance include decreased mental alertness and agility during the day, concentration problems, and mood fluctuations, often associated with temper outbursts. The sleep disturbance may even include bizarre and dangerous behavior during sleep. An example of the latter is REM sleep disorder.

About every ninety minutes during sleep, the EEG indicates alertness on the sleeper's part, and the sleeping person's eyes engage in rapid movements. Dreams occur during these rapid-eye-movement (REM) sleep episodes. But no matter how active dreams may be, the dreamer remains in bed and, with the exception of the rapid eye movements, gives no outward sign that he or she is dreaming. Not so with REM sleep disorder.

Typically a person afflicted with REM sleep disorder will awaken suddenly from sleep and within moments commit some impulsively violent act directed at himself or others. In one instance a man in his 60s described to me how he would suddenly awaken, leap from his bed, and rush straight into the bedroom wall. In other reported cases a person with the disorder has awakened from a sound sleep and started punching, kicking, and lashing out at anything or anyone within reach. Bed partners may be scratched, pummeled, or as in one famous case, strangled to death.

The explanation for these bizarre sleep-related events comes from the research with cats carried out by sleep researcher Michel Jouvet. Since cats spend a large portion of their sleep time in REM (the time when most dreams occur in humans), Jouvet reasoned that REM serves the same purpose for cats and that cats must be dreaming during this period. But how to get the cats to "tell" of their experiences? Jouvet provided an answer to this question by carrying out a meticulously controlled operation that severed the nerve pathways that extend from the cat's cerebral cortex downward to the spinal motor nerves. Normally during REM sleep these pathways convey inhibitory impulses originating in the cortex and terminating on the muscles. The result is paralysis of the cat during the REM period. This paralysis serves as a protective reflex: if the animal is pursued in its dream, it does not leap up to fight or flee.

When Jouvet's surgically modified cats entered REM sleep, they stood up, arched their backs, hissed at invisible attackers, or sniffed around their cages as if encountering them for the first time. These attack, defense, and exploratory behaviors bore no relationship to anything happening in the cat's vicinity but represented responses to internal events, that is, dreams. Since Jouvet's operation had reversed the usual paralysis attending REM, the cats acted out their dreams.

Recordings made of people waking from violent attacks confirm that, like the cats', their behavior occurs during emergence from REM sleep. Neuroscientists speculate that the attacks are due to a functional failure of the normal inhibitory pathways extending from brain to spinal motor nerve cells. A specific drug, Klonopin, ends the attacks by maintaining impulses along the inhibitory pathway.

Among the less dramatic examples of disturbed sleep that may occur in the mature brain are those resulting from depression, as well as from physical diseases. Think of insomnia as a *symptom* and, as with all symptoms, it is often only the outward sign of a deeper problem that can be discovered only by investigation. Thus insomnia is like recurrent headaches, which may be comparatively minor (tension headache), slightly more serious (migraine), or pernicious and life-threatening (a brain tumor). Sleep difficulties are also important to diagnose because they can lead to the overuse of sedatives and hypnotic medications.

An often unsuspected cause of sleep difficulties is the overuse of *alcohol.* While alcohol does induce drowsiness and lessens the time it takes to fall asleep, these beneficial effects are short lived. Rapid eye movement (REM) sleep and its associated dreams are greatly reduced or eliminated altogether. In the second half of the night, alcohol actively interferes with sleep and induces a restless turning and readjustment of position in an effort to get more comfortable. The resulting sleep is fitful and troubled by frequent awakenings.

Cigarette smoking leads to difficulty falling asleep and difficulty waking up (one of the reasons heavy smokers need that first "drag" in the morning in order to face the day). It's likely that sleep disturbances are more frequent among smokers due to a combination of several factors: the stimulant effects of nicotine, the nightly nicotine withdrawal during sleep, an increase of psychological disturbances and/or sleep-associated breathing disorders, such as increasing congestion in the mouth and throat and swelling in throat and upper air passages. Since smoking reduces the lungs' ability to take in oxygen from the air, this combination of reduced oxygen and reduced air passages leads to frequent awakening and chronic insomnia.

The brain-wave patterns of smokers show a decreased amount of deep sleep, which reverses if the smoker successfully kicks the habit. Among women, cigarette smoking often leads to excessive daytime drowsiness, while the same smoking habits in men typically produce nightmares and disturbing dreams.

In summary, with advancing years the mature brain undergoes alterations in its sleep needs. Typically, both falling asleep and remaining asleep become harder. It is unrealistic therefore for anyone over 50 to expect to experience the depth and quality of sleep enjoyed twenty or thirty years earlier. With age it is perfectly normal to awaken earlier, perhaps earlier than one wishes, and to feel somewhat less refreshed in the morning. It's useful to remember this fact whenever you find yourself staring at the ceiling while imagining your temporary insomnia as an indication of something seriously wrong. Nor are you helpless and unable to improve the quantity and quality of your sleep. Several measures can be taken to combat disturbed sleep. These include establishing regular hours for going to bed, eliminating lengthy daytime naps, avoiding alcohol late in the day, eliminating cigarettes, getting regular exercise during the day, and reducing the use of stimulants such as caffeine.

Sedatives and hypnotics are another approach to sleep problems. Nearly half of all hypnotic prescriptions written are for people over 65 years old. While sleeping pills may be helpful in the short run, they should not be used on a regular basis. Many pills disrupt the normal sleep cycle, and even those sleeping pills that do not have this side effect can decrease daytime alertness and impair memory. The affected person may appear sedated (as in fact he is), think and behave in a confused manner, fall, and even suffer from amnesia over things that have happened only a few hours before. The possibility for serious drug interactions is also greater if you are older. That's because people over 60 account for one-quarter of all the medicines prescribed in the United States. And since many of these people may be taking ten or more pills per day, any decrease in alertness or other cognitive dysfunction could lead to serious consequences.

An effective and easily obtainable sleep inducer is the natural brain hormone melatonin. The mature brain is unusually sensitive to supplemental melatonin. That's because with age the pineal gland within the brain decreases its level of melatonin production. A two-milligram controlled-release capsule will boost brain levels and effectively improve the quality and quantity of sleep without causing any significant side effects. If you take it you will fall asleep faster and spend more time in a deeper, more restful asleep. It can even be used for early-evening naps (6 P.M.–8 P.M.) since melatonin levels begin to fall during these hours.

Even better than any drug or supplement is the cultivation of certain habits and techniques that reduce worry:

- Go to bed at regular hours. Nothing is more conducive to regular restful sleep than a consistent sleep routine.

- Do not try to force sleep. Mental effort merely increases arousal, which is the greatest impediment to sleep. From this follows the next two rules.

- Do not exercise, attempt stressful work assignments, or engage in any other activity likely to bring into play the sympathetic nervous system: the brain's flight-fight component that has evolved to maximize alertness. And contrary to accepted wisdom in business circles, do not plan the events of your next day just before going to bed. This will only increase arousal and leave you mentally rehearsing for the future when you should be falling asleep.

- Avoid alcohol, tobacco, caffeine-containing substances, and if insomnia is a chronic problem, heavy meals taken after 5 P.M. (eat your main meal at lunch).

- If you can't fall asleep despite these measures, don't just lie there ruminating and fussing, but without looking at the clock (most people become more alert and aroused upon discovering that "it's already three o'clock and I'm still awake!"), get up and read something relaxing. Return to bed only when you feel sleepy. And while you may feel less rested the next morning as a result of some lost sleep, force yourself to remain on schedule and get up at the regular time in order to conform to the first rule.

The fourth mental health problem associated with the mature brain is *alcohol abuse*. Unfortunately, alcoholism is a common disease in older people, affecting up to 10 percent of those living at home and a startling 40 percent of those in nursing homes. While alcohol problems may be encountered across the life span, alcohol abuse later in life has its own specific features and effects. The same amount of alcohol that exerts only a mild effect on brain functioning in a person aged 30 through 55 may induce dramatic declines in functioning in an older person. That's because of the body's reduced ability to metabolize alcohol. As a result, as few as one or two drinks a day may increase disability and mortality.

Alcohol exerts its most disruptive effects on thinking, memory, orientation, and judgment. On occasion, alcohol-induced brain injury can resemble Alzheimer's disease. And the distinction between the conditions is important to make, since alcohol-related dementialike symptoms may largely reverse if the drinking is stopped.

Over-the-counter and prescription drugs interact with alcohol, greatly enhancing its effects. Older heavy drinkers often go to doctors (or are brought to them) because of associated (in medical terms, comorbid) mental health problems. Anxiety, depression, dementia, confusion, and sleep disturbances are the most frequent. Physical problems are also more common in heavy drinkers and include liver abnormalities, walking difficulties, and frequent falls.

Men and women also differ in the backgrounds leading up to their alcohol abuse. The men are more likely to be married, divorced, or separated, while women are more likely to be widows. In both sexes loneliness is the main contributing factor to heavy drinking. And since, as mentioned previously, adjustment to loneliness is the single most important determinant of a successful maturity, the possibility for late-onset alcoholism never entirely disappears. Loneliness also explains why some people who drank moderately or not at all during the first fifty-plus years of their lives develop alcoholism later in life. As one study concluded in regard to men, "The onset of problem drinking [later in life] has less to do with stress than with the loss of a spouse who may have served to regulate the older man's drinking."

Alcohol is not a total villain: it may even have positive effects if used in moderation. A study reported in the October 1996 issue of the *Journal of the American Geriatrics Society* concluded that a few glasses of beer or wine each week may provide intellectual benefits. Among the group of 2,040 people studied (average age 74), those who were light alcohol drinkers—consuming fewer than four drinks per week—exhibited slightly yet consistently higher scores on tests measuring mental function, compared to those who reported drinking ten or more drinks per week. Also in support of sensible and moderate alcohol use is a study over twenty years of the drinking habits of 4,739 twins. Comparisons of the health effects of moderate drinkers with nondrinkers failed to show in older individuals any association between moderate long-term alcohol drinking and lower cognitive performance. Instead there was even the suggestion of a protective effect on normal brain functioning related to past moderate alcohol drinking.

FOUR

ENHANCING BRAIN FUNCTION

WITH DRUGS

*N*euroscientists are developing medicines aimed at preventing brain diseases and brain damage, but they are also enthusiastically seeking agents that will enhance rather than just maintain brain function. In contrast to other brain-altering drugs, these cognitive enhancers (*nootropica*, as they are referred to technically) are intended not only to favorably influence the processes leading to various failures of cognition but, in addition, to improve brain function in people not known to be suffering from any brain disease (normals). These drugs are sometimes referred to as smart drugs or cognitive enhancers.

Neuroscientists are concentrating principally on nootropic drugs that will enhance brain functions such as learning and memory, thereby favorably influencing the aging process. These drugs will accomplish this by exerting a positive influence on one or more of the following age-related processes: the loss of cells and cell networks in the hippocampus; the decrease in neurotransmitter levels; imbalances between a host of different neurotransmitters.

Interestingly, several of the first generation of these compounds were accidentally discovered during drug trials aimed at restoring the memories of demented patients. To everyone's surprise, the drugs turned out to be of more help in improving the memories of normal volunteers than of individuals suffering from memory problems.

Several drugs currently on the market for unrelated illnesses are showing promise as memory aids. For instance, Indocin, an antiarthritis drug, was recently discovered to improve the short-term memory and coordination of healthy elderly volunteers. Other classes of drugs currently under investigation include vasodilators, anticoagulants, hormones, and compounds derived from brain tissue—referred to as neurotropic factors. Indeed the very number and variety of chemical approaches are an indicator that neuroscientists still do not have a handle on the basic mechanisms of cell senescence and death.

But such a variety of approaches isn't necessarily a bad thing. Our understanding of the mechanisms of brain disease has always lagged behind our penchant for devising new treatments for illnesses we don't understand. As I showed in my previous book, *Receptors*, most of the mind- and mood- altering drugs introduced into neuropsychiatry over the past three decades resulted from chance observations, lucky hunches that turned out to be correct. So it should be neither surprising nor disappointing that neuroscientists are seeking chemicals capable of altering aspects of brain aging prior to their achieving a consensus about the mechanisms involved.

At the moment, the development of nootropics is a kind of Holy Grail for several internationally based pharmaceutical companies that hope to be the first to develop a chemical capable of stopping or significantly delaying the development of dementia and the aging process. (Aging here is understood in a strictly functional rather than chronological point of view —as characterized by a progressive loss of complexity in all physiological systems and a gradual decrease in adaptability to the world.)

Nootropics work by enhancing normal brain functioning and thereby improving memory, concentration, psychological endurance, and other mental traits and activities.

The first commercially available nootropic, developed in the 1940s for the treatment of hypertension, was Hydergine. An extract of ergot, a fungus that grows on rye, Hydergine has FDA approval for the treatment of dementia and cerebro-vascular disease. According to the *Physicians Desk Reference*, the drug offers "some" relief from age-related declines in mental acuity. It is thought to do so by increasing metabolism and the utilization of glucose in the brain. Based on animal research, it is also thought likely that Hydergine helps rejuvenate nerve cells and protects the brain against damage due to lowered oxygen. By increasing the amounts of blood and oxygen to the brain, Hydergine checks the production of free radicals. While the drug is approved only for dementia, some physicians prescribe the drug as an aid to combating the effects of aging and improving intelligence. Mental alertness, clarity, and improved mood are among the benefits attributed to the drug.

While Hydergine is widely used in Europe, it is infrequently prescribed in the United States. Its lack of popularity here is at least partly due to a 1990 study from Jefferson Medical College in Philadelphia that concluded that Hydergine was ineffective as a treatment for Alzheimer's disease. This single study exerted a powerful influence on American pharmaceutical companies, which over the past decade have vied with one another to come up with a drug for the cure or arrest of Alzheimer's disease. In 1994, years of research along these lines culminated in the release of tacrine hydrochloride, the first drug approved by the FDA for Alzheimer's disease, sold under the trade name Cognex. This drug works by interfering with the breakdown of the neurotransmitter acetylcholine.

Of all the neurotransmitters, acetylcholine has received the most attention in drug development for dementia. This is because loss of cholinergic cells (cells that produce acetylcholine) is a consistent finding in the brains of people who have died from Alzheimer's disease. The loss is most evident in parts of the cortex and the hippocampus. This reduction in brain tissue capable of producing acetylcholine leads to a marked reduction in levels of that neurotransmitter in the brain. Tacrine hydrochloride counters this reduction by inhibiting the action of the enzyme acetylcholinesterase, which breaks acetylcholine down into simple chemicals.

As discussed in the previous chapter, neurotransmitters like acetylcholine are produced by the presynaptic neuron, which releases it into the synapse. After crossing the synapse and attaching to receptors on a receiving neuron (the postsynaptic neuron), the neurotransmitter is either taken back up by the presynaptic neuron or, as with acetylcholine, is broken down by the action of acetylcholinesterase. Since tacrine hydrochloride works by countering the action of this enzyme, it is classi-fied as a member of the class of drugs called cholinesterase inhibitors.

Memory function in humans is measured by standard memory tests that emphasize either language or drawings. After the subject has been told something or has read some-thing or, in nonverbal tasks, has looked at something such as a geometric drawing, he or she is asked about it. Subjects afflicted with Alzheimer's disease quickly forget what they have learned. Such failures are the result of the loss of cholin-ergic (acetylcholine) fibers that occurs in the disease.

In humans, the administration of a drug that inhibits the destruction of acetylcholine has been of only limited success in improving memory. Tacrine hydrochloride, for instance, helps only some patients and appears to have no effect on the progression of disease in the majority of patients. That's because memory in humans involves several neurotransmitter systems acting in concert. Correcting for cholinergic (acetyl-choline) deficiencies has no effect on deficiencies in other neu-rotransmitters. The situation is similar to what would happen if a new conductor were to attempt to enhance the perfor-mance of his orchestra by concentrating entirely upon the contributions of a single instrument.

Increased emphasis is now being placed on interactions between neurotransmitter systems. For instance, an approach that combines stimulation of both cholinergic and gluta-minergic (glutamine) systems is more likely to be successful than a concentration on either neurotransmitter system alone.

Another approach concentrates on the development of drugs capable of altering the concentration and distribution of monoamines, the group of neurotransmitters that includes dopamine, serotonin, and norepinephrine. Each of these messenger molecules is broken down by one of two paths. Either they are reabsorbed and repackaged by the presynaptic neuron in the process of reuptake or they are broken down into simpler chemical components, also within the presynaptic neuron, by the action of the enzyme monoamine oxidase (MAO).

The activity of MAO throughout the body helps to maintain stable neuronal firing rates within the brain. But this MAO-initiated breakdown of the monoamine neurotransmitters, particularly the breakdown caused by one of the two forms of MOA, MOA-B, generates toxic hydroxyl radicals. These radicals are members of the oxygen-associated free-radical group believed to be involved in neurodegenerative disorders like Parkinson's disease as well as normal aging. The relevance of this for drugs aimed at improving mental functioning in the later years?

As we age, the concentration of MAO-B increases, leading to deficiencies in the monoamine neurotransmitters and the accumulation of free radicals. This increase in MOA-B occurs in the brain stem, the limbic system, and the cerebral cortex, particularly the frontal cortex. Changes in any or all of these areas are most likely responsible for many of the mental changes associated with aging. Since MAO-B is principally responsible for these changes and the ensuing increase in free radicals, neuroscientists are actively pursuing the possibility that altering the levels of MAO-B might exert a protective effect on aging. One encouraging indicator is the positive effect brought about by lowering the levels of MAO-B in Parkinson's disease.

Several years ago neurologists introduced a MOA-B-inhibiting drug for Parkinson's disease called l-deprenyl (also called selegiline or Eldepryl). Parkinson's disease is thought to result from the toxic effect of free radicals acting on the nerve cells in the part of the brain stem (the substantia nigra) that manufactures dopamine. Drugs capable of decreasing the production of free radicals can therefore be expected to improve Parkinson's disease.

Inhibiting MAO-B with l-deprenyl exerts a protective action, presumably by decreasing the volume of free-radical production. Thus, Parkinson patients taking deprenyl experience a slowing of the progression of their disease as a result, it's thought, of inhibiting the action of MAO-B. Less MAO-B activity translates into fewer free-radical molecules, and this translates into less destruction of dopamine-producing cells in the brain stem.

Microscopic analysis of the brains of l-deprenyl–treated rats reveal a higher density of nerve fibers in the frontal cortex and hippocampus. The accumulation of lipofuscin, the aging pigment that clutters up the inside of neurons, was also reduced, suggesting a general slowing of age-associated brain changes. "Long-term treatment with l-deprenyl is able to counter some microanatomical changes typical of the aging frontal cortex and hippocampus of the aging rat," one Italian research group concluded (that is, l-deprenyl has proven capacity for reversing brain changes associated with aging).

L-deprenyl's actions, it turns out, are not limited to merely delaying the onset of a neurodegenerative disease like Parkinson's and countering the microscopic brain changes of aging. If the drug is given to rats at about 24 months of age, the animals' life expectancy is doubled.

Left at this, neuroscientists would probably have put the

life extension associated with l-deprenyl down to some statistical fluke and moved on to other projects. But the rats not only lived longer, they lived smarter. Among other things, they were again able to negotiate complicated mazes and escape from water tanks as well as younger rats. Similar improvements, though less reliable and less dramatic, occur in dogs given l-deprenyl.

So far nobody is absolutely certain why l-deprenyl enhances mental performance and opposes the aging process in the brain. Some neuroscientists believe it may be due to the induction and release of neuronal growth factors—general-purpose "tonics" that perk up nerve cells. Others believe prevention of free radicals resulting from the MAO-B inhibition does the trick. A final theory relates deprenyl's action to its induction of superoxide dismutase (SOD), which, you will recall, tends to be present in greater amounts in strains of species that live longer. But whatever the mechanism, l-deprenyl is the only drug that, up to this point, has halted, even reversed, aging in animal brains. So far, no one is certain if l-deprenyl can do the same for the human brain.

Lack of proven effectiveness has not halted a kind of bandwagon effect of enthusiasm for l-deprenyl. At a neurology meeting two years ago, a speaker extolling the value of l-deprenyl in Parkinson's disease asked his audience how many of the doctors were already taking or were seriously thinking of taking l-deprenyl as a life-extending agent. About a third of the neurologists in the auditorium raised a hand. Think of the implications of that: among a group of brain experts trained in the diagnosis and treatment of brain diseases, a third would be willing to apply to themselves results that so far have been obtained only in animals.

Is the neurologists' enthusiasm for l-deprenyl's life-extending properties a barometer of how the general population will respond when and if a retardant for brain aging is made available? I believe that it is. But before asking your assent to that, I should in all fairness tell you one more thing about l-deprenyl. It is rapidly metabolized in the body into forms of amphetamine and methamphetamine. Amphetamine and its chemical cousins are known for their energizing and mood-elevating effects. It's possible, therefore, that the doctors' enthusiasm for l-deprenyl is merely a reflection that the drug, via its amphetamine breakdown products, acts as an upper that simply makes them feel better.

• • •

Another widely touted nootropic is melatonin, mentioned earlier in regard to sleep. The ability of the brain's pineal gland to produce and release melatonin varies greatly with age. During the first few months after birth, the hormone is barely detectable in the blood. Thereafter, a melatonin rhythm develops rapidly. At night the pineal produces melatonin, and as a consequence, the highest levels occur at night, when it is dark. With age, melatonin production gradually decreases so that, in very old individuals of species ranging from algae to humans, the circadian (daily) melatonin rhythm is barely discernible. Neuroscientists speculate that perhaps the lessening of melatonin production with aging desynchronizes other circadian rhythms. If so, this desynchronization might contribute significantly to aging and render older individuals more susceptible to age-related diseases—at least that is the speculation.

Although some disagreement exists among neuroscientists

on the relationship between decreased melatonin production and aging, they are in complete agreement that aging of the pineal gland and preservation of the melatonin rhythm in animals can be achieved with food restriction. In rats fed 60 percent less food, the melatonin rhythm was equivalent to that in much younger animals. As mentioned earlier, food restriction is also a powerful longevity enhancer. Is there a relationship between the preserved melatonin rhythm and the increased longevity of food-restricted animals? At the moment there is no certain answer to that question.

Neuroscientists are in agreement, however, that melatonin is the most potent free-radical scavenger thus far discovered. It quenches what scientists consider the most toxic and damaging free radical produced in the body, the hydroxyl radical (OH).

As discussed in Chapter One, the most currently accepted theory of aging attributes the rate of aging to the effect of accumulated free-radical damage over the life span. Since melatonin can protect cellular molecules, especially DNA, against free-radical attack, its presence could be a major factor in determining the rate at which organisms age.

In addition to its ability to directly scavenge the highly toxic hydroxyl radical, melatonin promotes the activity of the antioxidative enzyme (glutathione peroxidase), which reduces oxidative reactions and their attendant damage.

If the anatomical and functional changes associated with aging are considered a consequence of accumulated free-radical damage, then a treatment aimed at altering the aging process immediately suggests itself: offset the normal age-associated drop in melatonin by the ingestion of supplemental melatonin. The additional melatonin might be especially helpful in the brain because that organ is highly susceptible to damage by oxygen-based radicals, damage that is particularly serious because of the brain's inability to regenerate destroyed or seriously damaged cells.

Further, supplemental melatonin may prove beneficial in delaying age-related degenerative diseases, many of which are also thought to be the result of free-radical oxidant damage.

"If we accept the premise that free radical damage that accumulates over a lifetime is consequential in aging and age-related diseases, it seems reasonable to assume that the pineal and melatonin may relate to these processes," writes Russel J. Reiter and his colleagues from the University of Texas Health Science Center in San Antonio. "Since melatonin is nontoxic, readily absorbed, when administered by any route . . . its utility as an antioxidant seems almost unlimited," they write in the *Annals of the New York Academy of Sciences* in 1996.

While this reasoning about melatonin as a longevity promoter seems sound enough, there is simply no way of proving it in long-lived species such as our own. Melatonin may make sense in the short term for sleep induction, but its long-term benefits for longevity prolongation have not been established.

In addition, a general principle applies here that relates to all supplements based on replacement for hormones or other chemicals that decline with aging.

"Maybe the melatonin level is falling as we age for some good reason that we don't know about," according to neuroscientist Richard Sprott. "By replacing melatonin and some of these other substances declining with age we may be inadvertently screwing things up in ways we don't yet know about."

Writing in the journal *Nature*, neurobiologist Fred Turek, a veteran of over twenty years of melatonin research, urges a cautious approach: "There has always been, and probably always will be, public enthusiasm for quick snake-oil cures to complex problems." While he considers melatonin a safe and effective sleep-inducing agent, he has considerable reservation about claims that melatonin prolongs life.

• • •

Another chemical touted to extend the life span is DHEA (dehydroepiandrosterone), the most plentiful steroid hormone in the human body. It is nicknamed the "mother steroid" because the body converts it into estrogen, testosterone, and other steroids. Like melatonin, DHEA levels wane with increasing age.

From childhood through early adulthood, the adrenal glands secrete DHEA in increasing amounts. From then on, the production of this critical hormone declines until, at about age 80, the amount of DHEA has been reduced to about 10 percent of its level at age 25.

Several studies suggest a positive effect of DHEA on longevity, vigor, and resistance to cancers and heart disease. These findings have stimulated enthusiasm in some scientific circles to investigate using DHEA in replacement therapy. In theory this should lead to life extension.

According to William Regelson, an immunologist at the Medical College of Virginia and coauthor of *The Melatonin Miracle*, DHEA and melatonin act in synergy by manipulating the body clock. According to this scenario, melatonin resets the clock and extends life while DHEA keeps the body free of disease. DHEA also inhibits oxygen free-radical formation. According to current thinking, these free radicals contribute to the development of many age-related diseases, including cancer, atherosclerosis, and various degenerative brain diseases.

Evidence gathered over the last decade indicates that the brain is capable of making DHEA and other *neurosteroids.* These are thought to play a key role in maintaining neuronal structure and growth, particularly in brain areas concerned with learning and memory. DHEA improves memory in aging mice. The same memory-improving effect in humans was reported in a 1995 study on the cognition-enhancing effects of DHEA in depression.

While all of this is highly suggestive that DHEA may function as a brain enhancer, such an interpretation suffers from several problems. For one, proponents of DHEA as an antiaging agent place great emphasis on its success in mice. When given the hormone, mice live longer, develop fewer cancers, and fight off infections with greater vigor—all thanks to a resurgence of their immune systems. But mice have little DHEA in their body at any time in their lives and may therefore respond to it differently than humans. In addition, the hormone can cause serious liver damage—in one study, fourteen of sixteen rats fed DHEA for eighty-four weeks developed liver cancer. Prompting additional concern is the fact that no one knows for certain how DHEA works. "How DHEA exerts its biologic actions remains an enigma," according to John E. Nestler of the Division of Endocrinology and Metabolism of the Medical College of Virginia at the

Virginia Commonwealth University in Richmond. "It seems likely that several independent mechanisms for DHEA action may be operative."

At this point, DHEA has shown promise in the treatment of autoimmune diseases like lupus, where the body attacks joints or skin or other of its own components. People taking the hormone report feeling and sleeping better while experiencing less pain and discomfort. But do such positive effects in a disease indicate that the hormone prolongs life? And if it did, how would anyone be able to prove it, at least in the short run?

With the average age of both men and women slowly creeping upward toward the early 80s, any significant effect on longevity would take almost a century to express itself. In the meantime, scientists have no other choice but to test potential longevity-enhancing agents on animals, like rats and mice, with short life spans. And even if the animals live longer, their life extension may have no relevance to human life spans. In other words, there is no way of conclusively proving in the short term the life-extending potential of any chemical in humans.

Nevertheless some researchers like Maria Dorota Majewska, of the Medications Development Division of the National Institute on Drug Abuse, suggest that use of DHEA may be advantageous for aging individuals. "A few controlled studies and a wealth of anecdotal evidence," says Majewska, "indicate that DHEA treatment in aging patients is both safe and beneficial. . . . While we are waiting for more complete data, the existing evidence already suggests that DHEA replacement may be a safe and effective means of improving health and the quality of life."

The operative word in Majewska's last sentence is "may." For most people, myself included, more certainty is required than the suggestion that a replacement therapy or any therapy for that matter "may be" effective.

Along with questions of efficacy is another important issue: Is DHEA safe enough to warrant its use in humans based simply on anecdote and animal experiments?

Not according to John E. Nestler, who coorganized a New York Academy of Sciences conference on DHEA and aging. Says Nestler:

As a physician I believe a few words of caution are in order. Some physicians are dispensing DHEA in cavalier fashion for almost any indication. This is not justified. Although the results of DHEA studies appear promising and tantalizing, they still need to be confirmed in large-scale and properly controlled studies. A beneficial effect of DHEA administration in humans has not yet been firmly established, and we know virtually nothing about the side effect profile of chronic DHEA administration. Without confirmed beneficial actions in humans and a better understanding of associated risks, it does not seem reasonable to dispense DHEA.

Over the last three decades, several other classes of drugs have shown promise in animals as cognitive enhancers but have not proven effective in humans. Extracts from the leaves of *Ginkgo biloba* improve the retention of learned behavior in rats. In humans, the herbs shorten the latency time needed to evaluate a particular stimulus and decide whether or not to react.

Another class of drugs, the piracetam nootropics, has been

under study now for more than thirty years. While the compounds are able to enhance memory in animals and reverse drug-induced amnesia, no commonly accepted mechanism of action has been established. In one trial of piracetam carried out in Lyons, France, the investigators cautiously suggested that "long-term administration of high doses of piracetam might slow the progression of cognitive deterioration in patients with Alzheimer's disease." The most significant finding among the piracetam group was a slight improvement in recent and remote memory.

Measurements of improvement in dementia are often taken as indicators that a new drug will prove helpful in people with normal brains. While this seems like a reasonable assumption, dementia, as we have stressed repeatedly throughout this book, is not simply an exaggerated form of normal aging. What may help the specific disease processes responsible for dementia may have little or no effect on the cognitive changes accompanying normal aging, and vice versa.

But even if we dispense with that important objection and consider a drug-induced improvement in dementia as an indicator that the same drug may work as a cognitive enhancer in normals, a problem remains. Dementia is often subtle and notoriously difficult to measure. For one thing, dementia may vary from day to day, and even with no treatment at all the affected person may on occasion seem much improved. This variability makes it difficult to say for certain whether any degree of improvement results from a nootropic or the observed result merely reflects the characteristic fluctuations of the illness.

• • •

A final approach to slowing the aging process involves the administration of drugs aimed specifically at neutralizing the effects of free-radical-induced damage. Several different scenarios seem possible. Perhaps a drug could be developed to interfere with the release of glutamate. Alternately, released glutamate could be neutralized by an antagonist. Or mitochondrial function could be boosted so that the calcium-ensnaring operation would not be accompanied by an overproduction of free radicals. Finally, the free radicals themselves could be set upon by scavengers or chemicals that lead to nerve-cell replacement (trophic factors).

Another chemical treatment may involve the restoration of age-related deficits in receptors. Age-related decreases in NMDA receptors occur in the cortex and hippocampus. Indeed, this decrease of NMDA-receptor density might be one of the causes of aging. Neuroscientists cautiously suggest that cognition might be improved by compensating for the receptor deficits associated with aging via the use of cognition-enhancing nootropics.

As you have no doubt noticed, several of these therapeutic approaches aimed at preventing free-radical damage are also being applied to the treatment of stroke. Similar approaches are being suggested for the treatment and/or prevention of Parkinson's disease, epilepsy, Alzheimer's disease, alcohol intoxication and withdrawal, and chronic-pain syndromes.

While such a panoply of possible benefits is encouraging, it also raises an important question: Is it possible that the emphasis on L-glutamate and free radicals is just a fad, that other theories will replace it, that neuroscientists a decade from now will have moved on to something completely different?

"Such an eventuality is extremely unlikely," according to Zaven Khachaturian. "Of course, the details will change with additional discoveries; but the basic elements will endure. Our future understanding of the aging brain will involve refinements not radical revisions of the excitotoxic theory."

Other neuroscientists are in agreement with Khachaturian. "Pharmacological manipulation of the excitatory amino acid receptors is likely to be of benefit in important and common diseases of the brain and nervous system," predicts R. J. Thomas, a specialist in the role of excitatory amino acids in health and disease. He suggests that the future lies with the identification of additional receptor subtypes and the mapping of their distribution patterns within the brain. This will provide neuroscientists with multiple approaches to both aging and various forms of brain damage.

• • •

Although not considered nootropics in the strict definition of the term, several readily available chemicals can alter brain function for the better. The most powerful and controversial agents are amphetamine and nicotine. In a society intent on stamping out drugs of abuse along with smoking, the suggestion that amphetamine and nicotine may be beneficial is certain to stir up resistance. Yet this is exactly what neuroscientists are finding.

"There are just a few drugs that do improve memory performance significantly beyond normal performance," according to Richard Thompson, professor of psychology and biological sciences at the University of Southern California. "Two of these drugs, neither of which is recommended, are nicotine and amphetamine. Both do produce significant improvement in both reasoning and memory, and perhaps attention."

The first suggestion of a positive role for nicotine in brain function evolved, as with much of the research in this area, from experiments with rats. Under ordinary conditions, an aged rat is severely limited in its ability to learn and remember new ways of wending its way through a radial maze. But a substantial enhancement in performance results if nicotine is given to the animal prior to the trial learning experiments. This improvement involves a nicotine-receptor-induced increase in generalized alertness and/or improvement in higher integrative functions.

If you want to put this in human terms, ask any regular cigarette smoker. He or she will describe to you smoking-induced increases in alertness, concentration, focus, memory, and general well-being. These mental changes occur because nicotine stimulates one or more nicotine receptors within the brain. (Incidentally, I don't smoke cigarettes and have no agenda about smoking by other people, so please don't take my comments as an argument in defense of smoking or a proposal that you should start smoking in order to improve your brain function.)

Why would nicotine exert such a powerful effect on brain activities? Within the brain nicotine mimics the neurotransmitter acetylcholine. It does this by acting at a specific acetyl-choline-receptor site. Within the brain, nicotine releases a wide variety of neurotransmitters, including endorphins (the body's own morphinelike chemicals implicated in pleasure), and dopamine (thought to be the important neurotransmitter for the brain's reward systems). Within the brain many different subtypes of the nicotine receptor exist. The most important ones are located within certain parts of the limbic system, the brain area that mediates the pleasurable response to certain frequently abused drugs such as amphetamine. Thus the similar effects of amphetamine and nicotine should come as no surprise, since they both stimulate the same parts of the brain.

This reward effect (general mood improvement and sense of well-being) partially accounts for the addicting properties of tobacco. The reinforcing properties of nicotine are thought to reflect both stimulation and desensitization (more nicotine required to get the same effect) of different nicotinic receptor populations.

If nicotine can create good effects (memory enhancement, mood improvement), why not try to stimulate those nicotinic receptors for these effects while bypassing the receptors responsible for nicotine's addicting qualities? Indeed, that is exactly what neuroscientists are attempting to do.

Already some have reported a slowing of the rates of progression of Alzheimer's disease via the administration of nicotine by injection or in pill form. But the research is tricky. Nobody knows just how many nicotine receptors exist in the brain or how to tease out from the crowd the specific ones responsible for the good effects. The goal now is additional understanding of the behavioral effects of the activation of discrete nicotine-receptor subtypes. This will provide needed information in designing drugs for memory loss and dementia that will work via selective nicotinic-receptor stimulation.

• • •

Caffeine is another commonly used chemical that exerts a powerful effect on the brain. At least half of the world's population drinks tea, while coffee is the second-most-traded commodity in the world. Per capita consumption of caffeine in the United States from all sources (tea, coffee, cocoa, chocolate, and cola) is about two hundred milligrams per day (between one and two cups of coffee).

Caffeine works by blocking a brain receptor for the neurotransmitter adenosine. Adenosine decreases the release of many other neurotransmitters in specific brain regions through actions at its principal receptor. Adenosine's action in the brain induces sedation, decreases seizures, reduces pain, and lowers anxiety. Caffeine counteracts these actions by blocking the adenosine receptor (hence caffeine is referred to as an adenosine-receptor *antagonist*).

The alertness you experience after you drink a caffeine-containing substance is the subjective correlate of the arousing effect of caffeine on the brain. With just two cups of coffee taken close together (about two hundred milligrams of caffeine), the cortex is activated, the electroencephalogram (EEG) shifts into an arousal pattern, and drowsiness and fatigue decrease. As a result, performance on a simple speeded task is faster and more accurate. This shorter reaction time under caffeine is thought to result from an increase in sustained attention and faster central motor processes— that is, it takes less time to respond to a given stimulus.

Caffeine's effects on physical performance are well established. If you ingest a caffeine-containing substance (coffee, tea, cola, and so on) prior to exercise, you can increase your endurance and performance. In one study, participants in a fifteen-hundred-meter swimming meet performed significantly faster and with less effort after taking caffeine. Moreover, this added boost occurred at levels of caffeine below the limit allowed by the International Olympic Committee. This raises "serious ethical issues regarding the use of caffeine to improve athletic performance," according to L. L. Spriet, writing in the *International Journal of Sports Nutrition.* Spriet suggests as one solution adding caffeine to the list of banned substances, thereby requiring athletes to abstain from caffeine ingestion forty-eight to seventy-two hours prior to competition. My point here is not to take a position on competitive sports (a topic somewhat far afield from our subject, the mature brain) but to underscore the physical-energy-boosting effects of what most of us do not ordinarily think of as a drug. "Like the amphetamines, but to a much smaller degree, caffeine prolongs the amount of time an individual can perform physically exhausting work," according to Vanderbilt University School of Medicine psychopharmacologist Oakley Ray.

Among both young and older individuals, caffeine produces similar energy boosts in the mental sphere as well. In psychological tests, caffeine raised energy levels and improved performance. Attention, problem solving, and delayed recall are particularly improved. In regard to mood, there are decreases in tenseness and increases in clearheadedness, happiness, and calmness. There is even some suggestive evidence that regular caffeine consumption may protect against suicide! (At least in women.)

In a study of over eighty-six thousand women conducted by researchers at Harvard Medical School, those who drank two or three cups of coffee per day had a 66 percent decreased risk of killing themselves over a ten-year period, compared with those who never drank coffee. The lead researcher on the study, Dr. Ichiro Kawachi, speculates that caffeine use may have produced improvements in mood and feelings of well-being (well-recognized effects of caffeine use) and led to a lessening of depression.

Does this mean that caffeine-containing products should be considered as nonprescription mood elevators, that they should be freely ingested on a regular basis? Although I think such a conclusion may be warranted, caution is advisable.

Given caffeine's stimulating effects on brain and performance, it would seem a near perfect antidote to the low energy states that often prevail during the mature years. But when taken in excess (the exact quantity differs from person to person), caffeine can produce a variety of unpleasant, even dangerous, symptoms. Nervousness, irritability, tremulousness, insomnia, and elevated pulse and blood pressure can result from too much. (Some studies also suggest an increase in heart attacks or coronary artery disease.) But these problems are seen predominantly in regular and heavy consumers of caffeine.

Taken in smaller amounts, caffeine can compensate for the 50 percent loss of neurons from the "juice machines": the nuclei beneath the cortex that modulate brain activity by producing and forwarding neurotransmitters upward from the brain stem. Think of the older brain as lacking the usual level of energizing chemicals. One way of correcting for that is to judiciously integrate the use of physical and mental stimulants such as caffeine and nicotine (in the form of pills, not tobacco).

"The best simple summary is that 150 to 300 mg of caffeine (two to three cups) offsets fatigue-induced performance decrements in both physical and mental tasks, and slows the development of boredom, and *may*, in a rested, interested individual, increase motor and mental efficiency above control levels," according to Oakley Ray.

• • •

While caffeine and nicotine are freely available, the other psychostimulants, amphetamine and Ritalin, require a prescription. Both drugs increase physical activity, increase the subjective sense of energy, and induce varying states of euphoria. In excess, these traits can be exaggerated to the point of a paranoid psychosis difficult to distinguish from schizophrenia. Amphetamine works by interacting with the monoamine group of neurotransmitters (dopamine and norepinephrine, principally) along the path from the brain stem upward to the limbic system (the mesolimbic pathway). Enhancement of dopamine neurotransmission in the mesolimbic pathway is fundamental to the reinforcing properties of the drug.

At the molecular level, amphetamine increases the amount of dopamine and norepinephrine released into the synapse and interferes with the clearing of these neurotransmitters from the synapse. When taken without medical supervision (or if prescribed by a physician unfamiliar with the drug), amphetamine can produce addiction or at least a severe dependence. But when taken in controlled amounts and under enlightened supervision, it can provide a needed energy boost by counteracting the lethargy and lack of energy that commonly afflict the mature brain. The drug also enhances short- and long-term memory performance. Moreover, the improved recall extends beyond what might be expected by the increase in general arousal and attention that also result from the drug.

Methylphenidate, or Ritalin, is a milder stimulant than amphetamine, with a potency between amphetamine and caffeine. Both Ritalin and amphetamine cause similar behaviors and subjective effects, but different neurochemical responses. Some people respond to one drug but not the other. Thus, if one stimulant is not effective, the other might well be.

None of these drugs (and they are all drugs) should be used to excess or on a daily basis. But with your doctor's approval, psychostimulants can be judiciously employed to restore attention, concentration, and energy to levels naturally encountered in individuals much younger. Unfortunately, many doctors hold outdated and mistaken notions about the proper role of psychostimulants. While it is true that all of these substances have the potential to create dependency in users, the chance of developing a major problem declines during the mature years. The mature brain is more likely to benefit from psychostimulants simply because these drugs can supply the very thing that the brain lacks as a consequence of aging: energy.

• • •

The next generation of nootropic drugs will work in novel ways, not as correctives for something wrong in the brain but, instead, chemical boosts for the improvement of memory function. These brain "tonics" will be combined with memory-training exercises since, it's turning out, nootropic drug treatment and memory-training work on different brain functions and hence operate synergistically. But it's likely to be several years before a safe, dependable drug will be widely available for rejuvenating the brain and reversing age-related cognitive changes.

THIRTY STEPS

YOU CAN TAKE TO

ENHANCE YOUR BRAIN IN

THE MATURE YEARS

As a result of the research I carried out for this book, including many interviews with people who are aging successfully, happily, and creatively, I've developed a series of Pearls that can be applied in order to keep the brain operating at its optimum. Physicians refer to Pearls as bits of applied wisdom that can aid in the understanding of diseases and the treatment of patients. Although aging is not a disease but a part of the normal life cycle, it does take some thoughtful preparation in order to be navigated successfully. The following nuggets of applied wisdom and practical measures are based on well-founded principles that are unlikely to change very much, whatever scientists may discover in the future about aging and the brain. These Pearls can and should be applied *now, whatever your age,* in order to achieve a richer, more fulfilling late adulthood marked by healthy brain functioning.

1. The mature brain is neither better nor worse than the brain in earlier years of development. It is just different. It is important therefore to understand and accommodate yourself to these differences.

2. So far, the only tried and true life-span-enhancing efforts are these: stop smoking; fasten automobile seat belts; engage in weight-bearing exercises daily. At the moment there are no foods, chemicals, or drugs that have been convincingly guaranteed to prolong life in humans.

3. Keep up good physical health. Maintain normal levels of blood pressure, blood sugar, and cholesterol. If medications are required to control these factors, take the drugs according to your doctor's prescription. Good control with or without medications lessens the risk of stroke, diabetes, and other conditions that compromise the supply of oxygen and blood sugar going to the brain.

4. Immobilization and physical inactivity are to be avoided at all costs. Some form of gentle stretching exercise is an excellent health promoter. As mentioned earlier, tai chi is an excellent conditioning activity to improve balance and mobility. Equally important are various weight-bearing exercises like weight lifting and walking. To be avoided as *first-time* activities later in life are potentially harmful sports activities that emphasize rapid, highly coordinated responses, such as downhill skiing, gymnastics, and most competitive ball games. Of course, any or all of these activities can be continued later in life if they have been part of one's activity and exercise pattern earlier in life.

5. Exercisers are fine for cardiovascular fitness but are not helpful for building up bone and preventing osteoporosis. Sitting on an Exercycle isn't nearly as effective as another cardiovascular-enhancing exercise done while standing, for example, rapid walking outdoors or along a treadmill.

6. Make a firm commitment to walking at least four hours a week. This will cut by 25 percent your chances of dying over the next four years, in comparison to those who don't walk. Although there are no proven effects on brain function from a walking regimen, such a program will help indirectly by the proven beneficial effect on the brain of increasing HDL ("good" cholesterol), lowering blood pressure, and reducing weight. In response to requests for guidelines on how long it should take to walk one mile, Dr. James Rippe, associate professor of medicine at the Tufts University School of Medicine, has drawn up the following table listing the time it should take a person to walk a mile at different ages. These are only suggested guidelines, not mandates. Your goals may be more or less ambitious according to your personal health status. When in doubt, consult your doctor. Also start slowly and work up gradually to the suggested level of performance.

7. The best single exercise that can be done anytime and without any special equipment involves nothing

Men 50 to 59	*15:36 to 17:00 minutes*
Men 60 to 69	*16:18 to 17:30 minutes*
Men 70 to 79	*20:00 to 21:48 minutes*
Women 50 to 59	*14:24 to 15:12 minutes*
Women 60 to 69	*15:12 to 16:18 minutes*
Women 70 to 79	*15:48 to 18:48 minutes*

more elaborate or complicated than standing on one foot for as long as possible and then switching to the other foot and doing the same thing. This seemingly simple exercise combines muscle strengthening, balance, and flexibility. The only requirements are (1) *start off slowly* (no more than fifteen or twenty seconds in the beginning) and (2) do it close to a wall, furniture, or some other *source of support* until you build up some experience with the exercise and become less likely to lose your balance and fall. I first learned of this exercise from discussions with Chinese doctors who told me that all patients in the hospitals in China who are capable of supporting their own weight are encouraged to get out of bed every day and stand by their bedside for three to five minutes. Those well enough and strong enough are then asked to do the weight-shifting part of the exercise. With time and practice this simple exercise can be increased to fifteen minutes at a time on one leg.

8. Reduce stress. The best way of doing this is by mentally reformulating everyday frustrations and problems into challenges. Think: What can I learn from this experience? Such reformulations give you more feeling of control; and the more you perceive yourself in control, the less likely unwanted experiences will be perceived as stressful. As a last resort, remember: even in an instance where you cannot change what's happening to you, you can always change your inner attitude toward it.

9. A slowing of the speed of general responsiveness is an inevitable result of aging. But this is not an occupational handicap, since experience can moderate the influence of slowing on work performance. In the professions such as law and medicine, rapidity of response plays almost no part: the courtroom and the consulting room provide plenty of time for the exercise of wisdom accumulated over a long career. The same holds true elsewhere in the workforce. An experienced waiter will usually outperform his younger counterpart because he has learned to conserve his physical and mental energy by means of informal memory aids and a nicely balanced attention to the needs of customers at his different tables. "Large affairs are not performed by muscle, speed, nimbleness, but by reflection, character, judgment. In age, these qualities are not diminished but augmented," wrote Cicero in *De Senectute*.

10. Physiological measurements confirm what older persons experience every day: decreased levels of energy in the mature years. Often the only difference between the brain functioning of youth and old age is the amount of energy required to generate enthusiasm and get oneself up and going. This may be subjectively experienced as a lack of enthusiasm: a feeling that a visit to a friend or attendance at an event isn't worth the effort. The good news is that the brain can be energized via the deliberate cultivation of curiosity. On the basis of my interviews and discussions, I have become convinced that curiosity is the mental trait most linked with superior brain functioning over the life span. The mentally healthy person of whatever age is deeply interested and curious about the people and events around him or her. Earlier in life such curiosity often comes naturally; in the later years it must often be deliberately cultivated. Read over again the comments of Art Buchwald and Olga Hirshhorn about the importance of curiosity and involvement.

11. A related point about energy: caffeine and other stimulating drugs (available by prescription) can make more energy resources available in times of decreased enthusiasm and energy reserve. We often forget that caffeine is an energizing drug that alerts the brain, improves concentration, and provides renewed energy. Amphetamine is another energy booster that in moderation and under a physician's direction can counteract the effects of lowered neurotransmitter levels in the brain's subcortical circuits, the "juice machines."

One more point about energy: learn the art of napping for short periods during the day. Contrary to standard but mistaken medical advice, napping does not interfere with nighttime sleep. Sleep researchers have discovered that naps do not serve to compensate for unmet sleep needs but, instead, provide a temporary respite from the day's activities and lead to improvements in energy, alertness, and mood.

12. Anyone over 50 years of age regularly experiences momentary lapses, when a name or specific word is just "on the tip of the tongue" and yet cannot be recalled until later. Such experiences are no cause for concern. Moreover, fretting about this perfectly normal age-associated memory impairment (AAMI) only makes things worse, since anxiety also interferes with recall. Healthy adaptations to AAMI include reading and utilizing the principles espoused in a memory training book such as *The Memory Book* by Harry Lorrayne and Jerry Lucas. Another useful tool is palmtop computers. They are small enough to be carried in a pocketbook and yet are powerful enough to hold several volumes' worth of information. (These include the Wizard or Taurus by Sharpe, the Hewlett-Packard Palmtop PCs, and my personal favorite, the Psion 3a by Psion Inc.)

13. Even in the presence of some degree of memory impairment, dementia is unlikely. A diagnosis of dementia requires impairments in memory plus at least one more brain function, such as language, thinking, or conceptualization. So don't waste time and energy worrying that you are becoming senile. Of course, if memory problems are interfering with your life or leading to conflicts with other people, you should seek consultation with a neurologist. Reversible causes for memory failures exist and often can be successfully treated.

14. Rather than worrying about dementia, take specific steps to make it less likely. One way to do this is to keep working as long as possible (assuming you are reasonably satisfied with your occupation or profession). In addition, avoid the six factors shown to be associated with increased likelihood of dementia: low levels of physical activity, failure to retain a high degree of finger dexterity, less-frequent opportunities to converse, too much empty spare time, a decreased number of friends, excessive use of alcohol.

15. Your attitudes and activities over the next five minutes can exert more of an influence on your brain power than your genetic inheritance. You can enhance brain performance and health by choosing to stimulate and challenge your brain: pick up that newspaper and do that crossword puzzle now.

16. Keep a diary of your daily activities, your thoughts, your daytime fantasies, and your nighttime dreams. This not only is another continuity enhancer but serves as a means of learning the patterns and habits that govern your life. What did you do and what were your thoughts during that blizzard last January? Six months later, while sitting in the sun on your deck reading your diary entries (perhaps interspersed with some snapshots you took at the time), you will enjoy not only pleasure but needed linkages of present to past. Essentially, the "old age of the mind" consists in meaninglessness and the absence of continuity. You can prevent meaninglessness by using your diary. If writing doesn't appeal to you, you can also use one of the palmtops mentioned in Pearl 12 instead. If you do so, however, be sure to protect your privacy by not leaving the palmtop lying around for prying eyes to see. Also, regularly back up your entries by downloading the information on one of the solid-state discs that can be purchased for each model.

17. Take the same attitude toward your social life as you do toward your investments: diversify rather than invest everything in a single area. Gerontologist Gene Cohen suggests that mature people should develop a "social portfolio" divisible into four areas. Included under *active group activities* are such things as dance and tennis, which involve high energy and high mobility. *Passive group activities* are low-energy and low-velocity and include things like volunteerism, art classes, and game clubs. *Individual active activities* are also high-energy and/or high-mobility, such as nature walks or jogging. *Individual passive activities* are low-energy and low-mobility and include reading, writing letters, cooking, or working on a crossword puzzle. Since social isolation and loneliness can exert such destructive effects as one grows older, participation in both active and passive group activities becomes particularly important.

18. If you're not computer literate, take some lessons that provide you with the necessary background to get on the Internet. Most newspapers print classified ads by computer specialists offering basic instructions. While the Internet is not a substitute for real-life social relations, it can counter feelings of loneliness and isolation. "Going on-line allows you to be intellectually mobile and be socially mobile," says Mary Furlong, founder of a group called SeniorNet. As an example of Ms. Furlong's point, an elderly widower who had played bridge with his wife for the forty years before her death joined a bridge group on the Internet. Now he's playing with enthusiasts all over the world. Next winter he is signed up for a bridge cruise in the Caribbean, where he will meet many of his new friends face-to-face. For him the Net provided not only a means of keeping his mind focused and sharp but an opportunity to form new friends in the process.

19. Seek out opportunities for sensory stimulation. The maintenance of visual and other sensory acuity is linked with retained physiological and psychological integrity of the brain. Go to the museum and look for extended periods at a picture by Monet. What was he seeing when painting water lilies? Or listen to a CD recording of Glenn Gould playing the *Goldberg Variations.* Figure out for yourself what Gould meant when he described this work as a "vision of subconscious design exulting upon a pinnacle of potency."

In addition, take adult education courses in new and unfamiliar subjects whenever possible. They are likely to be as protective of healthy brain functioning in the mature years as the formal education attained earlier in life. In one study of centenarians, the formal average educational level was only 5.8 years. There is no reason for discouragement, therefore, if you ended your education prematurely. Education is a lifetime activity, and the benefits don't stop at the conclusion of high school or college.

20. With age, the brain suffers a loss in its capacity for sustained concentration, with the upper limit being about fifteen minutes. This curtailed concentration time is not abnormal. Concentration difficulties are encountered elsewhere along the life span (for example, in children, particularly children with hyperactivity). Nor does a shortened concentration time preclude any type of intellectual pursuit. What it does require is that as you get older you divide your periods of active mental work into fifteen-minute sessions broken up by a distraction period of between three and five minutes. During these breaks some people simply close their eyes and rest. Others turn to something routine that doesn't require any concentration, such as a phone call to place an order. Others find some type of short physical activity best, such as walking to the corner to mail a letter. The important thing is that any period of sustained mental activity be briefly interrupted every fifteen minutes.

21. When facing mental challenges, go slowly, check your work, draw on your years of experience, and rely less on your speed of response. Reaction time lengthens with age. But that small liability can almost always be compensated for by wisdom and accumulated life experience.

22. Games like bridge, chess, and bingo help maintain sharpness in different mental domains. Bridge and chess involve working memory, reasoning, attention, and timing. As mentioned earlier, no age-related differences were found on measures of thinking and memory in bingo players—a strong recommendation for bingo as a means of providing positive benefits for older players. As mentioned in Pearl 20, focus and concentration often suffer because of a greater susceptibility to distraction. Playing chess, bridge, or bingo on a regular basis can counter this tendency, since all three of these games involve situations that call for sustained concentration under conditions of high arousal.

23. Try to retain a sense of humor. Humor scores highly among traits contributing to longevity. Taking oneself too seriously not only leads to unfortunate consequences when it comes to friends (no one likes spending time with people who lack a sense of humor) but leads to increased illness, disability, and a premature death.

24. Do everything you can to keep up your present friendships and strike up new ones. The brain is an inherently social structure that works on the microscopic scale via the interaction of billions of connections. On the macroscopic scale, we are the mirror images of that connectivity and flourish best when we interact meaningfully and emotionally with other people. Remember the advice of Arlene, who recovered from her depression by following these rules: "Take an interest in other people and keep abreast of what is going on in their world.

Exercise and go out for a walk every day, even if it's only for a little while. Try to have a number of friends, and try to be a good friend. Keep up your physical appearance, because if you feel you look good, you will have more confidence and will be better able to meet people and make new friends."

Make special efforts to keep up your contacts with younger people. This is important because it provides continuity with the issues that occupied your earlier years (work, child care, housing, the quality of the schools). Your sense of humor is particularly important here since some younger people may, out of their own fears of aging, try to keep you at a distance.

As a correlate and help to fulfilling this Pearl, try to maintain a nonjudgmental attitude toward the people and events around you. This doesn't mean relinquishing moral and ethical principles. Rather, a nonjudgmental attitude prevents you from prematurely narrowing your interests. As one older college professor told me during our interview, "So many people decide what interests them and what they can and should do, and therefore close their minds off to things that may teach them a lot." Remember: life is a full-time, extended learning experience. Don't cheat yourself by prejudging what is and is not worth learning.

25. Loneliness is the greatest challenge to be overcome as you advance toward the mature years. Loss and a certain degree of isolation are inevitable. Rather than denying loneliness, try to build up your tolerance for being alone. This doesn't contradict the recommendations of Pearl 24 about social involvement, but complements them: only by being comfortable with yourself and finding pleasure in your own company can you be pleasurable company for someone else.

If you live alone, consider getting a pet. Pets can do a lot for assuaging the pangs of loneliness and reducing the risk of depression. Although there are more pet cats than dogs in the nation (63 million cats, 54.2 million dogs), a dog might be a better choice. For some reason a dog confers greater protection against loneliness than a cat—perhaps because dogs have to be walked, and walking brings the owner into contact with other dog owners. Another factor: physical exertion and exercise combat feelings of loneliness, and dog owners report taking twice as many daily walks as nonowners. According to a study done at the School of Veterinary Medicine, University of California, dog owners voice less dissatisfaction with their social, physical, and emotional states. But whatever pet you choose, interaction with the animal is likely to decrease your blood pressure and pulse rate. Pet owners also are more likely to be alive one year after a heart attack than people who do not keep a pet. My guess is that these health benefits from pet ownership result from the socialization that results from owner-pet interaction. Almost all pet owners talk discursively to their pet, confide in their pet, and generally feel bonded to the pet. Conversations with other pet owners consist almost entirely of pet-related topics—further enhancing the socialization among pets and pet owners. The end result of all of this is a decrease in isolation and an enhancement of the limbic system and other brain circuits involved in emotion.

26. Diet. Your best chance for increasing your functional life span (period of healthy productive life) is by reducing the damage done by free radicals. This can be accomplished by (*a*) moderate caloric restriction, (*b*) decreased eating of foods that tend to increase free-radical reaction levels, for example, copper, polyunsaturated fats, or easily oxidized amino acids, (*c*) increasing your ingestion of foods containing effective natural free-radical reaction inhibitors (fruits and vegetables like cabbage, cauliflower, and carrots).

Radical caloric restriction does not make sense as an antidote to aging. What's best is a nutritionally balanced diet that enables you to maintain your weight at about the average for age and height.

Vitamins are probably not necessary to a healthy mental functioning, and some of them may in fact be harmful. Fresh foods contain vitamins in shelf-stable and shelf-unstable forms naturally balanced against one another. But only the shelf-stable forms are put in vitamin supplements, thus skewing the natural balance. Research by Dr. Victor Herbert, professor of medicine at Mount Sinai School of Medicine in New York, shows that ingesting only one form of nutrient is not beneficial—and can even worsen general health by triggering heart disease, cancer, and liver and kidney damage. Nor do the vitamin supplements provide the many other micronutrients present in food substances that may in fact be just as important for good health as the vitamins themselves. The best approach is to gain needed nutrients from food —at least five servings daily of fruits and vegetables. (One serving equals one piece of large fruit or about one-half cup of vegetables or berries.) If you insist on vitamins, then stick to recommended doses of vitamin pills containing inhibitors of free-radical reactions (vitamins E, A, and C).

Zinc deficiency is common with increasing age. This deficiency leads to disturbances in several zinc-dependent functions such as wound healing, skin-cell turnover, taste acuity, and immune efficiency. These are all factors that as a rule are depressed in older people. In many instances, these age-dependent falloffs in performance can be reversed by zinc supplements. Several over-the-counter vitamins contain zinc in amounts well in excess of minimum daily requirements.

Dietary calcium should be taken by women starting in their twenties. Osteoporosis is a leading killer and disabler of women in their later years: 50 percent of women who enter a nursing home with a broken hip never leave that nursing home alive. If osteoporosis can be delayed three years by lifetime dietary calcium supplements, the life span of many women can be extended into the eighties.

27. If you are a woman and have entered menopause, supplementary estrogen should probably be taken. Exceptions may exist for women who come from families with a history of breast and uterine cancer. Estrogen not only serves as a protectant against heart and other degenerative diseases but, scientists suspect, can prevent the onset of dementing illnesses like Alzheimer's.

28. Melatonin may make sense for nighttime sleep disturbances, but no convincing research exists that it exerts any positive effect on longevity. In regard to vitamins and longevity, no evidence exists that vitamin E helps; beta-carotene may be harmful; vitamin C exerts positive effects on general health but not necessarily longevity.

29. DHEA looks promising as an antiaging agent, and I urge you to follow research findings over the next several years. Health benefits in humans have already been demonstrated, and so far no overwhelmingly negative side effects have emerged. But as endocrinologist John Nestler mentions, it may not be reasonable to take DHEA now. But if you're over 65 you may be willing to take the risk. As William Regelson, the foremost proponent of DHEA, puts it, "I'm 70 years old. I don't want to wait twenty or thirty years. I want it *now*."

30. Alcohol can be taken in moderation. But monitor very closely how much you are actually drinking. As discussed earlier, alcoholism can develop late in life even among people who had no problem with alcohol when they were younger. Never drink when alone, before the late afternoon or early evening, and never as a means of combating loneliness, anger, or other uncomfortable emotions.

The Bottom Line

Aging can be thought of as the result throughout the body of a general wear-and-tear process. In all body organs except the brain, increased activity leads to more wear and tear and accelerated degeneration. But in the brain, the principle of operation is unique. Activation of nerve cells doesn't lead to a general degeneration of function but, instead, to the maintenance of neurons during normal aging. This is really quite an extraordinary situation if you think about it: the brain, in contrast to every other organ in the body, has the potential to *improve* with use and to keep that edge into the ninth decade and beyond.

Why does the brain work better the more it is used? Some neuroscientists believe this enhanced performance results because neuronal stimulation induces protective mechanisms such as DNA repair within neurons. Others emphasize the role of brain circuits: the connections between neurons. Each time the circuit is used, the connections become stronger and more easily facilitated. New activities and learning experiences establish new circuits and favor the survival of all of the neurons involved in the circuit. But whatever the mechanism, this "Use it or lose it" principle helps explain why nerve cells degenerate as a result of disuse in unhealthy aging but continue to function normally in robust aging. This same principle explains why recovery of various neuronal systems during aging can often be achieved by renewed stimulation. The situation is no different from what happens during physical training: muscular exercise results in increases in mass and strength, whereas inactivity leads to atrophy and muscle weakness.

Moreover, activation of brain cells and circuits provides a means of prolonging the brain's optimal functioning for the full length of our natural life span. From this it's fair to conclude that, barring health problems, the state of our brain during the mature years determines how long we will live. The Dutch Longitudinal Study Among the Elderly followed 211 Dutch people aged 65–84 years over an eight-year period. In those 70 years or older, the greater the decline in mental functioning, the shorter the life span. The investigators concluded that the "rate of decline of cognitive function is an independent predictor of longevity in older persons." This rule is valid even among people with Alzheimer's disease: the worse the mental impairment, the shorter the life span. Thus all of us face the same challenge: if we fail to employ our brain in varied and challenging ways, our very survival is threatened.

• • •

The take-home message when it comes to the brain? *Your brain: use it or lose it.* Indeed, it was my discovery of the intimate interplay of brain health, enhanced mental functioning, and longevity that stimulated me to write this book. After my research and interviews I became convinced—and I hope by now I have convinced you—that when it comes to longevity the brain is not just one organ among many but *the pivotal determiner of how long we will live.* Thus, providing our brain with challenges and stimulation is not only desirable but mandatory if we wish to enjoy a long and healthy life.

BIBLIOGRAPHY

Amenta, F.; Bongrani, S.; Cadel, S.; Ricci, A.; Valsecchi, B.; Zeng, Y. C. 1994. Neuroanatomy of the aging brain. Influence of treatment with L-deprenyl. *Ann. N.Y. Acad. Sci.* 717 (June 30): 33–44.

Arendash, G. W.; Sanberg, P. R.; Sengstock, G. J. 1995. Nicotine enhances the learning and memory of aged rats. *Pharmacol. Biochem. Behav.* 52, no. 3 (Nov.): 517–23.

Bachmann, G. Q. 1995. Influence of menopause on sexuality. *Int. J. Fertil. Menopausal Stud.* 40, Suppl. 1: 16–22.

Baltes, P. B.; Dittmann-Kohli, F.; Kliegl, R. 1986. Reserve capacity of the elderly in aging-sensitive tests of fluid intelligence: Replication and extension. *Psychol. Aging* 1, no. 2 (June): 172–77.

Ban, T. A. Psychopharmacology and successful cerebral aging. 1995. *Prog. Neuropsychopharmacol. Biol. Psychiatry* 19, no. 1 (Jan.): 1–9.

Bellino, F. L.; Daynes, R. A.; Hornsby, P. J.; Lavrin, D. H.; Nestler, J. E. 1995. Dehydroepiandrosterone (DHEA) and aging. *Ann. N.Y. Acad. Sci.*, vol. 774: 1–350.

Bilger, B. 1995. Forever young: Can DHEA temper the ravages of time? Some investigators can't afford to wait for the answer. *The Sciences*, Sept.-Oct., 26–31.

Brown, D. R.; Wang, Y.; Ward, A.; Ebbeling, C. B.; Fortlage, L.; Puleo, E.; Benson, H.; Rippe, J. M. 1995. Chronic psychological effects of exercise and exercise plus cognitive strategies. *Med. Sci. Sports Exerc.* 27, no. 5 (May): 765–75.

Bruce-Jones, P. N.; Crome, P.; Kalra, L. 1994. Indomethacin and cognitive function in healthy elderly volunteers. *Br. J. Clin. Pharmacol.* 38, no. 1 (July): 45–51.

Butler, R. N.; Lewis, M. I. 1995. Late-life depression: When and how to intervene. *Geriatrics* 50, no. 8 (Aug.): 44–46, 49–52, 55, 56–57 (quiz).

Casey, D. A. 1994. Depression in the elderly. *South. Med. J.* 87, no. 5 (May): 559–63.

Casper, R. C. 1995. Nutrition and its relationship to aging. *Exp. Gerontol.* 30, nos. 3-4 (May-Aug.): 299–314.

Chodzko-Zajko, W. J.; Schuler, P.; Solomon, J.; Heinl, B.; Ellis, N. R. 1992. The influence of physical fitness on automatic and effortful memory changes in aging. *Int. J. Aging Hum. Dev.* 35 (4): 265–85.

Christensen, H.; Henderson, A. S. 1991. Is age kinder to the initially more able? A study of eminent scientists and academics. *Psychol. Med.* 21, no. 4 (Nov.): 935–46.

Christensen, H.; Mackinnon, A.; Jorm, A. F.; Henderson, A. S.; Scott, L. R.; Korten, A. E. 1994. Age differences and interindividual variation in cognition in community-dwelling elderly. *Psychol. Aging* 9, no. 3 (Sept.): 381–90.

Christian, J. E.; Reed, T.; Carmelli, D.; Page, W. F.; Norton, J. A., Jr.; Breitner, J. C. 1995. Self-reported alcohol intake and cogniton in aging twins. *J. Stud. Alcohol* 56, no. 4 (July): 414–16.

Clarkson-Smith, L.; Hartley, A. A. 1990. The game of bridge as an exercise in working memory and reasoning. *J. Gerontol.* 45, no. 6 (Nov.): 233–38.

Clayton, G. M.; Martin, P.; Poon, L. W.; Lawhorn, L. A.; Avery, E. L. 1993. Survivors of the century. *Nurs. Health Care* 14, no. 5 (May): 256–60.

Cohen, P. 1996. Greying population stays in the pink. *New Scientist*, Mar. 16, 4.

Cole, T. R.; Winkler, M. G. 1994. *The Oxford book of aging.* New York: Oxford University Press.

Colerick, E. J. 1985. Stamina in later life. *Soc. Sci. Med.* 21 (9): 997–1006.

Commissaris, C. J.; Verhey, F. R., Jr.; Ponds, R. W.; Jolles, J.; Kok, G. J. 1994. Public education about normal forgetfulness and dementia: Importance and effects. *Patient Educ. Couns.* 24, no. 2 (Oct.): 109–15.

Corpas, E.; Harman, S. M.; Blackman, M. R. 1993. Human growth hormone and human aging. *Endocr. Rev.* 14, no. 1 (Feb.): 20–39.

Daffner, K. R.; Scinto, L. F.; Weintraub, S.; Guinessey J. E.; Mesulam, M. M. 1992. Diminished curiosity in patients with probable Alzheimer's disease as measured by exploratory eye movements. *Neurology* 42, no. 2 (Feb.): 320–28.

Daffner, K. R.; Scinto, L. F.; Weintraub, S.; Guinessey J. E.; Mesulam, M. M. 1994. The impact of aging on curiosity as measured by exploratory eye movements. *Arch. Neurol.* 51, no. 4 (Apr.): 368–76.

Deberdt, W. 1994. Interaction between psychological and pharmacological treatment in cognitive impairment. *Life Sci.* 55 (25-26): 2057–66.

Deeg, D. J.; Hofman, A.; van Zonneveld, R. J. 1990. The association between change in cognitive function and longevity in Dutch elderly. *Am. J. Epidemiol.* 132, no. 5 (Nov.): 973–82.

Denney, N. W. 1995. Critical thinking during the adult years: Has the developmental function changed over the last four decades? *Exp. Aging Res.* 21, no. 2 (Apr.-June): 191–207.

D'Esposito, M.; Detre, J. A.; Alsop, D. C.; Shin, R. K.; Atlas, S.; Grossman, M. 1995. The neural basis of the central executive system of working memory. *Nature* 378, no. 6554 (Nov.): 279–81.

Disterhoft, J. F.; Gispen, W. H.; Traber, J.; Khachaturian, Z. S. 1994. Calcium hypothesis of aging and dementia. *Ann. N.Y. Acad. Sci.*, vol. 747: 1–482.

Bibliography

Deptula, D.; Singh, R.; Pomara, N. 1993. Aging, emotional states, and memory. *Am. J. Psychiatry* 150, no. 3 (Mar.): 429–34.

Erber, J. T.; Rothberg, S. T. 1991. Here's looking at you: The relative effect of age and attractiveness on judgements about memory failure. *J. Gerontol.* 46, no. 3 (May): 116–23.

Erber, J. T.; Szuchman, L. T.; Rothberg, S. T. 1990. Everyday memory failure: Age differences in appraisal and attribution. *Psychol. Aging* 5, no. 2 (June): 236–41.

Evans, W. J. 1995. Effects of exercise on body composition and functional capacity of the elderly. *J. Gerontol. A. Biol. Sci. Med. Sci.* 50, spec. no. (Nov.): 147–50.

Fabrigoule, C.; Letenneur, L.; Dartigues, J. F.; Zarrouk, M.; Commenges, D.; Barberger-Gateau, P. 1995. Social and leisure activities and risk of dementia: A prospective longitudinal study. *J. Am. Geriatr. Soc.* 43, no. 5 (May): 485–90.

Farragher, B.; Wrigley, M.; Veluri, R. 1994. Alcohol related problems in elderly people—A prospective study. *Ir. Med. J.* 87, no. 6 (Nov.-Dec.): 172–73.

Ferrucci, L.; Guralnik, J. M.; Marchionni, N.; Costanzo, S.; Lamponi, M.; Baroni, A. 1993. Relationship between health status, fluid intelligence and disability in a non-demented elderly population. *Aging* (Milano) 5, no. 6 (Dec.): 435–43.

Finfgeld, D. L. 1995. Becoming and being courageous in the chronically ill elderly. *Issues Ment. Health Nurs.* 16, no. 1 (Jan.-Feb.): 1–11.

Fulop, T., Jr.; Seres, I. 1994. Age-related changes in signal transduction. Implications for neuronal transmission and potential for drug intervention. *Drugs Aging* 5, no. 5 (Nov.): 366–90.

Garfinkel, D.; Laudon, M.; Nof, D.; Zisapel, N. 1995. Improvement of sleep quality in elderly people by controlled-release melatonin. *Lancet* 346, no. 8974 (Aug. 26): 541–44.

Garvey, M. J.; Schaffer, C. B. 1994. Are some symptoms of depression age dependent? *J. Affect. Discord.* 32, no. 4 (Dec.): 247–51.

Gorman, W. F.; Campbell, C. D. 1995. Mental acuity of the normal elderly. *J. Okla. State Med. Assoc.* 88, no. 3 (Mar.): 119–23.

Gottfries, C. G. 1990. Neurochemical aspects of aging and diseases with cognitive impairment. *J. Neurosci. Res.* 27, no. 4 (Dec.): 541–47.

Gouliaev, A. H.; Senning, A. 1994. Piracetam and other structurally related nootropics. *Brain Res. Rev.* 19, no. 2 (May): 180–222.

Grady, C. L.; McIntosh, A. R.; Horwitz, B.; Maisog, J. M.; Ungerleider, L. G.; Mentis, M. J.; Pietrini, P.; Schapiro, M. B.; Haxby, J. V. 1995. Age-related reductions in human recognition memory due to impaired encoding. *Science*, 269 (July 14): 218–20.

Grimby, A. 1993. Bereavement among elderly people: Grief reactions, post-bereavement hallucinations and quality of life. *Acta Psychiatr. Scand.* 87, no. 1 (Jan.): 72–80.

Bibliography

Haimov, I.; Lavie, P.; Laudon, M.; Herer, P.; Vigder, C.; Zisapel, N. 1995. Melatonin replacement therapy of elderly insomniacs. *Sleep* 18, no. 7 (Sept.): 598–603.

Heikkinen, M. E.; Isometsa, E. T.; Aro, H. M.; Sarna, S. J.; Lonnqvist, J. K. 1995. Age-related variation in recent life events preceding suicide. *J. Nerv. Ment. Dis.* 183, no. 5 (May): 325–31.

Heseker, H.; Schneider, R. 1994. Requirement and supply of vitamin C, E and beta-carotene for elderly men and women. *Eur. J. Clin. Nutr.* 48, no. 2 (Feb.): 118–27.

Homma, S.; Ishida, H.; Hirose, N.; Nikamura, Y. 1994. Direct mail investigation of the social and physical background of centenarians in Tokyo metropolitan area (in Japanese). *Nippon Ronen Igakkai Zasshi* 31, no. 5 (May): 380–87.

Howieson, D. B.; Holm, L. A.; Kaye, J. A.; Oken, B. S.; Howieson, J. 1993. Neurologic function in the optimally healthy oldest old. Neuropsychological evaluation. *Neurology* 43, no. 10 (Oct.): 1882–86.

Inagaki, T.; Yamamoto, T.; Niimi, T.; Hashizume, Y.; Mizuno, T.; Inagaki, A.; Ojika, K. 1995. A 115-year-old woman: The oldest individual in Japan (in Japanese). *Nippon Ronen Igakkai Zasshi* 32, no. 3 (Mar.): 172–77.

Israel, L.; Melac, M.; Milinkevitch, D.; Dubos, G. 1994. Drug therapy and memory training programs: A double-blind randomized trial of general practice patients with age-associated memory impairment. *Int. Psychogeriatr.* 6, no. 2 (fall): 155–70.

Johnston, J. E. 1994. Sleep problems in the elderly. *J. Am. Acad. Nurse. Pract.* 6, no. 4 (Apr.): 161–66.

Jones, S. J.; Rabbitt, P. M. 1994. Effects of age on the ability to remember common and rare proper names. *Q. J. Exp. Psychol.* [A] 47, no. 4 (Nov.): 1001–14.

Jutagir, R. 1994. Psychological aspects of aging: When does memory loss signal dementia? *Geriatrics* 49, no. 3 (Mar.): 45–46, 49–51, 52–53 (quiz).

Kallinen, M.; Markku, A. 1995. Aging, physical activity and sports injuries. An overview of common sports injuries in the elderly. *Sports Med.* 20, no. 1 (July): 41–52.

Kaye, J. A.; Grady, C. L.; Haxby, J. V.; Moore, A.; Friedland, R. P. 1990. Plasticity in the aging brain. Reversibility of anatomic, metabolic, and cognitive deficits in normal-pressure hydrocephalus following shunt surgery. *Arch. Neurol.* 47, no. 12 (Dec.): 1336–41.

Kenemans, J. L.; Lorist, M. M. 1995. Caffeine and selective visual processing. *Pharmacol. Biochem. Behav.* 52, no. 3 (Nov.): 461–71.

Kitani, K.; Aoba, A.; Goto, S. 1996. Pharmacological intervention in aging and age-associated disorders. *Ann. N.Y. Acad. Sci.*, vol. 786: 1–460.

Lai, J. S.; Lan, C.; Wong, M. K.; Teng, S. H. 1995. Two-year trends in cardiorespiratory function among older tai chi chuan practitioners and sedentary subjects. *J. Am. Geriatr. Soc.* 43, no. 11 (Nov.): 1222–27.

Bibliography

Lai, J. S.; Wong, M. K.; Lan, C.; Chong, C. K.; Lien, I. N. 1993. Cardiorespiratory responses of tai chi chuan practitioners and sedentary subjects during cycle ergometry. *J. Formos. Med. Assoc.* 92, no. 10 (Oct.): 894–99.

Lang, F. R.; Carstensen, L. L. 1994. Close emotional relationships in late life: Further support for proactive aging in the social domain. *Psychol. Aging* 9, no. 2 (June): 315–24.

Lehr, U. 1991. Centenarian—A contribution to longevity research (in German). *Z. Gerontol.* 24, no. 5 (Sept.-Oct.): 227–232.

Levy, B.; Langer, E. 1994. Aging free from negative stereotypes: Successful memory in China and among the American deaf. *J. Pers. Soc. Psychol.* 66, no. 6 (June): 989–97.

Lindenberger, U.; Baltes, P. B. 1994. Sensory functioning and intelligence in old age: A strong connection. *Psychol. Aging* 9, no. 3 (Sept.): 339–55.

Lissner, L.; Bengtsson, C.; Bjorkelund, C.; Wedel, H. 1996. Physical activity levels and changes in relation to longevity. A prospective study of Swedish women. *Am. J. Epidemiol.* 143, no. 1 (Jan.): 54–62.

MacRae, P. G.; Asplund, L. A.; Schnelle, J. F.; Ouslander, J. G.; Abrahamse, A.; Morris, C. 1996. A walking program for nursing home residents: Effects on walk endurance, physical activity, mobility, and quality of life. *J. Am. Geriatr. Soc.* 44, no. 2 (Feb.): 175–80.

Mann, D. 1996. Depression among the elderly often overlooked, untreated. *Medical Tribune News Service*, May 8.

Bibliography

Mari, D.; Maestroni, Am. 1995. Centenarians in Milan (in Italian). *Ann. Ital. Med. Int.* 10, no. 1 (Jan.-Mar.): 46–48.

Martin, A. J.; Friston, K. J.; Colebatch, J. G.; Frackowiak, R. S. 1991. Decreases in regional cerebral blood flow with normal aging. *J. Cereb. Blood Flow Metab.* 11, no. 4 (July): 684–89.

Masoro, E. J. 1993. Dietary restriction and aging. *J. Am. Geriatr. Soc.* 41, no. 9 (Sept.): 994–99.

McEntee, W. J.; Crook, T. H. 1991. Serotonin, memory, and the aging brain. *Psychopharmacology* (Berlin) 103 (2): 143–94.

McEntee, W. J.; Crook, T. H. 1993. Glutamate: Its role in learning, memory, and the aging brain. *Psychopharmacology* (Berlin) 111 (4): 391–401.

Medina, J. J. 1996. *The clock of ages: Why we age—How we age—Winding back the clock.* New York: Cambridge University Press.

Mills, M. A.; Coleman, P. G. 1994. Nostalgic memories in dementia—A case study. *Int. J. Aging Hum. Dev.* 38 (3): 203–19.

Molander, B.; Backman, L. 1994. Attention and performance in miniature golf across the life span. *J. Gerontol.* 49, no. 2 (Mar.): 35–41.

Mondadori, C. 1993. The pharmacology of the nootropics: New insights and new questions. *Behav. Brain Res.* 59, nos. 1-2 (Dec. 31): 1–9.

Mondadori, C. 1994. In search of the mechanism of action of the nootropics: New insights and potential clinical implications. *Life Sci.* 55 (25–26): 217–87.

Montagu, J. D. 1994. Length of life in the ancient world: A controlled study. *J. R. Soc. Med.* 87, no. 1 (Jan.): 25–26.

Morrell, R. W.; Park, D. C.; Poon, L. W. 1990. Effects of labeling techniques on memory and comprehension of prescription information in young and old adults. *J. Gerontol.* 45, no. 4 (July): 166–72.

Mortel, K. F.; Meyer, J. S.; Herod, B.; Thornby, J. 1995. Education and occupation as risk factors for dementias of the Alzheimer and ischemic vascular types. *Dementia* 6, no. 1 (Jan.-Feb.): 55–62.

Okamoto, K.; Ohno, Y. 1994. Sociomedical and life-style risk factors of senile dementia, determined in nested case-control study (in Japanese). *Nippon Ronen Igakkai Zasshi* 31, no. 8 (Aug.): 604–9.

Petkov, V. D.; Kehayov, R.; Belcheva, S.; Konstantinova, E.; Petkov, V. V.; Getova, D.; Markovska, V. 1993. Memory effects of standardized extracts of Panax ginseng (G115), Ginkgo biloba (GK 501) and their combination Gincosan (PHL-00701). *Planta Med.* 59, no. 2 (Apr.): 106–14.

Pierpaoli, W.; Regelson, W.; Fabris, N. 1994. The aging clock: The pineal gland and other pacemakers in the progression of aging and carcinogenesis. *Ann. N.Y. Acad. Sci.*, vol. 719: 1–588.

Plassman, B. L.; Welsh, K. A.; Helms, M.; Brandt, J.; Page, W. F.; Breitner, J. C. 1995. Intelligence and education as predictors of cognitive state in late life: A 50-year follow-up. *Neurology* 45, no. 8 (Aug.): 1446–50.

Pokras, R. S. 1994. A possible role of glutamate in the aging process. *Med. Hypotheses* 42, no. 4 (Apr.): 253–56.

Rabbitt, P. 1990. Mild hearing loss can cause apparent memory failures which increase with age and reduce with IQ. *Acta Otolaryngol. Suppl.* (Stockholm) 476: 167–75; discussion, 176.

Receputo, G.; Rapisarda, R.; Motta, L. 1995. Centenarians: Health status and life conditions (in Italian). *Ann. Ital. Med. Int.* 10, no. 1 (Jan.-Mar.): 41–45.

Reiter, R. J.; Melchiorri, D.; Sewerynek, E.; Poeggeler, B.; Barlow-Walden, L.; Chuang, J.; Ortiz, G. G.; Acuna-Castroviejo, D. 1995. A review of the evidence supporting melatonin's role as an antioxidant. *J. Pineal Res.* 18, no. 1 (Jan.): 1–11.

Rich, J. B.; Rasmusson, D. X.; Folstein, M. F.; Carson, K. A.; Kawas, C.; Brandt, J. 1995. Nonsteroidal anti-inflammatory drugs in Alzheimer's disease. *Neurology* 45, no. 1 (Jan.): 51–55.

Ricklefs, R. E.; Finch, C. E. 1995. *Aging: A natural history.* New York: Scientific American Library.

Ritchie, K. 1995. Mental status examination of an exceptional case of longevity: J. C., aged 118 years. *Br. J. Psychiatry* 166, no. 2 (Feb.): 229–35.

Bibliography

Rogers, J.; Hart, L. A.; Boltz, R. P. 1993. The role of pet dogs in casual conversations of elderly adults. *J. Soc. Psychol.* 133, no. 3 (June): 265–77.

Rogers, P. G.; Richardson, N. J.; Dernoncourte, C. 1995. Caffeine use: Is there a net benefit for mood and psychomotor performance? *Neuropsychobiology* 31 (4): 195–99.

Rolls, B. J.; Dimeo, K. A.; Shide, D. J. 1995. Age-related impairments in the regulation of food intake. *Am. J. Clin. Nutr.* 62, no. 5 (Nov.): 923–31.

Rowland, D. L.; Greenleaf, W. J.; Dorfman, L. J.; Davidson, J. M. 1993. Aging and sexual function in men. *Arch. Sex. Behav.* 22, no. 6 (Dec.): 545–57.

Sadow, T. F.; Rubin, R. T. 1992. Effects of hypothalamic peptides on the aging brain. *Psychoneuroendocrinology* 17, no. 4 (Aug.): 293–314.

Salthouse, T. A. 1994. Age-related differences in basic cognitive processes: Implications for work. *Exp. Aging Res.* 20, no. 4 (Oct.-Dec.): 249–55.

Salthouse, T. A. 1996. General and specific speed mediation of adult age differences in memory. *J. Gerontol. B. Psychol. Sci. Soc. Sci.* 51, no. 1 (Jan.): 30–42.

Schiavi, R. C.; Rehman, J. 1995. Sexuality and aging. *Urol. Clin. North. Am.* 22, no. 4 (Nov.): 711–26.

Schmidt-Nielsen, K. 1994. About curiosity and being inquisitive. *Annu. Rev. Physiol.* 56: 1–12.

Small, G. W.; LaRue, A.; Komo, S.; Kaplan, A.; Mandelkern, M. A. 1995. Predictors of cognitive change in middle-aged and older adults with memory loss. *Am. J. Psychiatry* 152, no. 12 (Dec.): 1757–64.

Smith, M. C.; DeFrates-Densch, N.; Schrader, T. O.; Crone, S. F.; Davis, D.; Pumo, D. J.; Runne, J. T.; Van Loon, P. C. 1994. Age and skill differences in adaptive competence. *Int. J. Aging Hum. Dev.* 39 (2): 121–36.

Soetens, E.; Casaer, S.; D'Hooge, R.; Hueting, J. E. 1995. Effect of amphetamine on long-term retention of verbal material. *Psychopharmacology* (Berlin) 119, no. 2 (May): 155–62.

Spriet, L. L. 1995. Caffeine and performance. *Int. J. Sport. Nutr.* 5, Suppl. (June): S84–99.

Storandt, M. 1991. Memory-skills training for older adults. *Nebr. Symp. Motiv.* 39: 39–62.

Swaab, D. F. 1991. Brain aging and Alzheimer's disease, "wear and tear" versus "use it or lose it." *Neurobiol. Aging* 12, no. 4 (July-Aug.): 317–24.

Thomas, R. J. 1995. Excitatory amino acids in health and disease. *J. Am. Geriatr. Soc.* 43, no. 11 (Nov.): 1279–89.

Thomas-Anterion, C.; Marion, A.; Laurent, B.; Laporte-Simitsidis, S.; Foyatier-Michel, N.; Michel, D. 1994. Aging and procedural memory. Study of a series of tests on microcomputers (in French). *Rev. Med. Interne* 15 (9): 581–88.

Tierney, M. C.; Szalai, J. B.; Snow, W. G.; Fisher, R. H.; Tsuda, T.; Chi, H.; McLachlan, D. R.; St. George-Hyslop, P. H. 1996.

A prospective study of the clinical utility of ApoE genotype in the prediction of outcome in patients with memory impairment. *Neurol.* 46: 149–54.

Verhaeghen, P.; Marcoen, A.; Goossens, L. 1993. Facts and fiction about memory aging: A quantitative integration of research findings. *J. Gerontol.* 48, no. 4 (July): 157–71.

Warburton, D. M. 1995. Effects of caffeine on cognition and mood without caffeine abstinence. *Psychopharmacology* (Berlin) 119, no. 1 (May): 66–70.

West, M. J.; Coleman, P. D.; Flood, D. G.; Tronsoco, J. C. 1995. Differential neuronal loss in the hippocampus in normal aging and in patients with Alzheimer's disease. *Ugeskr. Læger* (in Danish) 157, no. 22 (May): 3190–93.

Wetter, D. W.; Young, T. B. 1994. The relation between cigarette smoking and sleep disturbance. *Prev. Med.* 23, no. 3 (May): 328–34.

Wolfson, L.; Whipple, R.; Judge, J.; Amerman, P.; Derby, C.; King, M. 1993. Training balance and strength in the elderly to improve function. *J. Am. Geriatr. Soc.* 41, no. 3 (Mar.): 341–43.

Woodward, W. 1991. New surprises in very old places: Civil War nurse leaders and longevity. *Nurs. Forum* 26 (1): 9–16.

Yoder, M. A.; Haude, R. H. 1995. Sense of humor and longevity: Older adults' self-ratings compared with ratings for deceased siblings. *Psychol. Rep.* 76, no. 3, pt. 1 (June): 945–46.

Bibliography

Zasloff, R. L.; Kidd, A. H. 1994. Loneliness and pet owner-ship among single women. *Psychol. Rep.* 75, no. 2 (Oct.): 747–52.

Zec, R. F. 1995. The neuropsychology of aging. *Exp. Geron-tol.* 30, nos. 3-4 (May-Aug.): 431–42.

Zisook, S.; Shuchter, S. R.; Sledge, P.; Mulvihill, M. 1993. Aging and bereavement. *J. Geriatr. Psychiatry. Neurol.* 6, no. 3 (July-Sept.): 137–43.

Zs.-Nagy, I.; Harman, D.; Kitani, K. 1994. Pharmacology of aging processes. Methods of assessment and potential inter-ventions. *Ann. N.Y. Acad. Sci.*, vol. 717: 1–350.

INDEX

Printed in the United States
By Bookmasters